# Unmanned and Autonomous Ships

# Unmanned and Autonomous Ships
## An Overview of MASS

R. Glenn Wright

LONDON AND NEW YORK

First published 2020 by Routledge

2 Park Square, Milton Park, Abingdon, Oxon OX14 4RN
605 Third Avenue, New York, NY 10017

*Routledge is an imprint of the Taylor & Francis Group, an informa business*

First issued in paperback 2022

Copyright © 2020 by Taylor & Francis Group, LLC

All rights reserved. No part of this book may be reprinted or reproduced or utilized in any form or by any electronic, mechanical, or other means, now known or hereafter invented, including photocopying and recording, or in any information storage or retrieval system, without permission in writing from the publishers.

Notice:
Product or corporate names may be trademarks or registered trademarks, and are used only for identification and explanation without intent to infringe.

Publisher's Note

The publisher has gone to great lengths to ensure the quality of this reprint but points out that some imperfections in the original copies may be apparent.

*British Library Cataloguing-in-Publication Data*
A catalogue record for this book is available from the British Library

*Library of Congress Cataloging-in-Publication Data*
A catalog record has been requested for this book

ISBN: 978-1-138-32488-6 (hbk)
ISBN: 978-1-03-233676-3 (pbk)
DOI: 10.1201/9780429450655

Typeset in Sabon
by Deanta Global Publishing Services, Chennai, India

To Magdalena …

# Contents

*Preface*   xvii
*Acronyms and Abbreviations*   xix

**1 Introduction**   1
   *1.1 A Historical Perspective on Advances in Shipping 2*
      *1.1.1 Construction 3*
      *1.1.2 Function 4*
      *1.1.3 Propulsion 4*
      *1.1.4 Navigation 5*
      *1.1.5 Communications 5*
      *1.1.6 Electronic Navigation and Communications 6*
      *1.1.7 Command and Control 8*
   *1.2 Current Initiatives in Unmanned and Autonomous Shipping 9*
      *1.2.1 Industry and Academia 9*
      *1.2.2 Regulatory Authorities 12*
      *1.2.3 Classification Societies 12*
      *1.2.4 Non-Governmental Organizations 13*
   *1.3 A Mariner's Perspective 13*
      *1.3.1 Human Senses Exceed Remote Operator and Full Autonomy Capabilities 14*
      *1.3.2 Implicit Devaluation of the Maritime Professions 15*
      *1.3.3 Maritime Jobs 16*
      *1.3.4 Other Issues 17*
   *References 17*

**2 Making the Case for Unmanned and Autonomous Ships**   21
   *2.1 Economic Perspectives 23*
      *2.1.1 Economical Transport 23*
         *2.1.1.1 Fixed Costs 24*
         *2.1.1.2 Operating Costs 25*

viii   Contents

   2.1.2 New Business Opportunities  26
   2.1.3 Data Monetization  27
 2.2 Safety  28
   2.2.1 Safety of Navigation  28
   2.2.2 Job Safety  29
     2.2.2.1 Hazard and Failure Prognostics and Detection  29
     2.2.2.2 Automation of Hazardous Tasks  30
 2.3 Environment  32
 References  33

## 3  Autonomy, Automation and Reasoning  37

 3.1 Metrics of Autonomy  37
   3.1.1 Lloyds Register  38
   3.1.2 Norwegian University of Science and Technology (NTNU)  40
   3.1.3 Norwegian Forum for Autonomous Ships (NFAS)  40
   3.1.4 Maritime UK  41
   3.1.5 Society of Automotive Engineers (SAE)  42
   3.1.6 IMO Definition of Autonomy  42
   3.1.7 Comparisons between Different Approaches to Autonomy  42
 3.2 Process Automation  44
   3.2.1 Propulsion Control  44
   3.2.2 Power Generation, Distribution and Control  45
   3.2.3 Auxiliary Systems  46
   3.2.4 Navigation  46
   3.2.5 Communications  47
   3.2.6 Alarm Monitoring and Damage Control  48
   3.2.7 Integration of Process Automation  49
 3.3 MASS Reasoning  49
   3.3.1 Some Thoughts on Reasoning  50
   3.3.2 Artificial Intelligence  51
     3.3.2.1 Neural Network Architecture  52
     3.3.2.2 Software Simulation  54
     3.3.2.3 Network Training  54
     3.3.2.4 Hardware Implementation  56
 References  57

## 4  MASS Design and Engineering  59

 4.1 Applications and Operational Settings  59
   4.1.1 Hull and Deck Design  60

- 4.1.2 Propulsion and Power Generation 61
  - 4.1.2.1 Engines 61
  - 4.1.2.2 Fuel and Power Sources 62
- 4.1.3 Sensors 64
  - 4.1.3.1 Environment Visualization 64
  - 4.1.3.2 Situational Awareness and Comprehension 65
  - 4.1.3.3 Sensor Suite Composition and Placement 65
- 4.1.4 Maintenance 69
  - 4.1.4.1 Introduction of Highly Reliable Systems 69
  - 4.1.4.2 Multiple Redundant Systems 70
  - 4.1.4.3 Predictive Maintenance 72
  - 4.1.4.4 Robotic Maintenance 73
- 4.2 Implementations of MASS 74
  - 4.2.1 Container and Bulk Shipping 74
    - 4.2.1.1 Yara Birkeland 74
    - 4.2.1.2 Project SeaShuttle 75
    - 4.2.1.3 Great Intelligence 75
  - 4.2.2 Ferries 76
    - 4.2.2.1 Folgefonn 76
    - 4.2.2.2 Falco 76
    - 4.2.2.3 Suomenlinna II 77
  - 4.2.3 Surveillance, Firefighting, Survey, and Search and Rescue 77
    - 4.2.3.1 Sharktech 77
    - 4.2.3.2 Sea Machines 78
    - 4.2.3.3 C-Worker 7 78
  - 4.2.4 Offshore Support 78
    - 4.2.4.1 Hrönn 78
    - 4.2.4.2 SeaZip 3 79
    - 4.2.4.3 Autonomous Spaceport Drone Ships (SpaceX) 79
  - 4.2.5 Tugboat 79
    - 4.2.5.1 RAmora 2400 79
    - 4.2.5.2 Svitzer Hermod 80
    - 4.2.5.3 Keppel Singmarine 80
- 4.3 Harbor Enhancements to Accommodate MASS 81
  - 4.3.1 Port and Harbor Facilities 81
    - 4.3.1.1 Automatic Berthing and Unberthing 81
    - 4.3.1.2 Bunkering 82
  - 4.3.2 Harbor Approaches 83
- References 83

## 5 Remote Control Centers 87

- 5.1 Transition to Remote Control and Supervision 88
- 5.2 Remote Control Center Functions 89
  - 5.2.1 Distribution of Authority 91
    - 5.2.1.1 Individual Ship Operations 91
    - 5.2.1.2 RCC Handling of Multiple Ship Operations 95
- 5.3 Remote Control Center Facilities 96
  - 5.3.1 Management Level 97
    - 5.3.1.1 Corporate Representative 97
    - 5.3.1.2 Captain and First Mate 99
    - 5.3.1.3 Chief Engineer 101
  - 5.3.2 Operational Level 103
    - 5.3.2.1 Second and Third Mates 103
    - 5.3.2.2 Helmsman 103
    - 5.3.2.3 Second and Third Engineers 106
    - 5.3.2.4 Security Officer 106
    - 5.3.2.5 Communications Officer 108
    - 5.3.2.6 Automation Officer 108
    - 5.3.2.7 Unmanned Air Vehicle (UAV)/ Unmanned Underwater Vehicle (UUV) Operator 109
  - 5.3.3 Support Level 109
    - 5.3.3.1 Technical Experts 109
    - 5.3.3.2 Other Support Staff 110
- 5.4 Remote Control Center Organization 110
  - 5.4.1 Functional Organization 110
  - 5.4.2 Allocation of Physical Space 112
  - 5.4.3 Computational Facilities 112
- References 113

## 6 Navigation 115

- 6.1 Aids to Navigation 116
- 6.2 Collision Avoidance 117
- 6.3 Environmental Sensor Systems 118
  - 6.3.1 Conventional vs. Smart Sensors 119
  - 6.3.2 Shipboard Sensors 120
    - 6.3.2.1 Surface Sensors and Systems 123
    - 6.3.2.2 Subsea Sensors 127
  - 6.3.3 Air-Based Sensors 129

        6.3.4    Space-Based Sensors  129
                 6.3.4.1   Global Navigation Satellite
                               System (GNSS)  129
                 6.3.4.2   Automatic Identification System (AIS)  130
                 6.3.4.3   Meteorological and
                               Oceanographic (METOC)  130
                 6.3.4.4   Other Sensors  131
        6.3.5    Sensor Data Types and Characteristics  131
        6.3.6    Sensor System Limitations and Vulnerabilities  133
                 6.3.6.1   GNSS Outages, Spoofing,
                               Jamming and Denial of Service  133
                 6.3.6.2   AIS Range, Clutter,
                               Spoofing and Jamming  134
                 6.3.6.3   Database Hacking  135
                 6.3.6.4   Multiple Sensor Modalities  136
  6.4    Navigational Reasoning  137
        6.4.1    Navigation under Nominal Conditions  138
        6.4.2    Navigation Absent Expected ATON  139
        6.4.3    Navigation under Threat of
                Collision, Allision or Attack  140
        6.4.4    Navigation under GNSS/AIS Denial
                of Service Attack or Spoofing  140
        6.4.5    Digital Twin Simulation  141
References  141

# 7  Communications    145

  7.1    General Communication Requirements for Ships  145
  7.2    e-Navigation Enhancements to Communications  146
  7.3    Limitations of e-Navigation as Relate to MASS  147
  7.4    Communications Requirements for MASS  148
  7.5    Communication Internal to MASS  149
        7.5.1    Navigation Systems  149
        7.5.2    Engineering Systems  150
        7.5.3    Imaging Systems  151
        7.5.4    Local Area Networks  151
  7.6    Bridge-to-Bridge Communication  152
        7.6.1    Lights and Shapes  153
        7.6.2    Sound and Light Signals  154
        7.6.3    Signals to Attract Attention and Distress Signals  154
        7.6.4    VHF Radio  155
        7.6.5    Distress Radio Communications  155

7.7 Communication between Ship and Shore 156
    7.7.1 Medium/High Frequency Radio 156
    7.7.2 Cellular Communication 157
    7.7.3 Satellite Communication 157
    7.7.4 Microwave Communication 157
7.8 MASS Area Communications 158
References 159

# 8 Security 161

8.1 It Begins with the Vessel Security Plan 161
8.2 MASS Ability to Maintain Security 162
    8.2.1 Qualifications 163
    8.2.2 Capabilities 164
8.3 Physical Security 164
    8.3.1 Unauthorized Entry 164
        8.3.1.1 Area Defenses 165
        8.3.1.2 Perimeter Defenses 166
        8.3.1.3 Exterior Defenses 167
        8.3.1.4 Internal Defenses 168
        8.3.1.5 Non-lethal Defenses 168
    8.3.2 Physical Attack 169
8.4 Threats Internal to the Vessel 169
    8.4.1 Proper Vetting of Authorized Personnel 169
    8.4.2 Obsolete Software 170
    8.4.3 Crew and Vendor Awareness 171
8.5 External Electronic Threats 171
    8.5.1 GNSS Spoofing and Denial of Service 172
    8.5.2 AIS Limitations 172
8.6 Cyber Security 173
    8.6.1 Incidents 173
    8.6.2 Implications for MASS 174
    8.6.3 Cyber Security Program Implementation 174
References 175

# 9 Training for MASS Operations 179

9.1 Technological Change 179
9.2 Training Curricula 181
    9.2.1 Bachelor of Science Degree considering Autonomous Shipping 181
    9.2.2 Master Degree with a Focus on Autonomous Shipping 182
    9.2.3 Maritime Training Centers 183

9.3  Licensing Requirements  183
    9.3.1  Licensing Endorsements  184
    9.3.2  Unlicensed Ratings  185
    9.3.3  Vessel Security Officer (VSO)  186
    9.3.4  Communications Officer  186
    9.3.5  Automation Officer  187
    9.3.6  Maritime Drone Operator  187
References  188

## 10  Regulatory Issues   189

10.1  International Maritime Organization (IMO)  190
    10.1.1  Regulatory Scoping Exercise  192
    10.1.2  Interim Guidelines for MASS Trials  193
10.2  Nation States  193
    10.2.1  European Union  194
    10.2.2  Denmark  194
    10.2.3  Finland  195
    10.2.4  The Netherlands  196
    10.2.5  Norway  196
    10.2.6  United Kingdom  196
    10.2.7  China  197
    10.2.8  Singapore  197
    10.2.9  Japan  198
    10.2.10  United States  198
    10.2.11  Other Countries and Related Issues  199
10.3  Classification Societies  200
    10.3.1  American Bureau of Shipping (ABS)  200
    10.3.2  Bureau Veritas (BV)  200
    10.3.3  China Classification Society (CCS)  201
    10.3.4  Det Norske Veritas-Germanischer Lloyd (DNV-GL)  201
    10.3.5  International Association of Classification Societies Ltd. (IACS)  201
    10.3.6  Korean Register of Shipping (KR)  201
    10.3.7  Lloyd's Register of Shipping (LR)  202
10.4  Non-Governmental Organizations  202
    10.4.1  Baltic and International Maritime Council (BIMCO)  202
    10.4.2  Comité Maritime International (CMI)  203
    10.4.3  International Chamber of Shipping (ICS)  203
    10.4.4  International Federation of Shipmasters' Associations (IFSMA)  203

10.4.5 *International Group of Protection and Indemnity Clubs (IGP&I) 204*
10.4.6 *Institute of Marine Engineering, Science & Technology (IMarEST) and the International Marine Contractors Association (IMCA) 204*
10.4.7 *International Organization for Standardization (ISO) 204*
10.4.8 *International Transport Workers' Federation (ITF) 205*
10.4.9 *Nautical Institute (NI) 205*
10.4.10 *One Sea Autonomous Maritime Ecosystem 205*
10.4.11 *Smart Ships Coalition (SSC) 206*
10.4.12 *Unmanned Cargo Ship Development Alliance 206*
References 206

## 11 Legal Issues    213

11.1 *Instruments Requiring Amendments to Support MASS Operations 214*
    11.1.1 *International Convention for the Safety of Life at Sea (SOLAS Convention) 215*
    11.1.2 *International Ship and Port Facility Security (ISPS Code) 221*
    11.1.3 *International Safety Management (ISM Code) 222*
    11.1.4 *International Convention on Standards of Training, Certification and Watchkeeping for Seafarers (STCW Convention) 222*
    11.1.5 *Convention on the International Regulations for Preventing Collisions at Sea (COLREG) 223*
    11.1.6 *International Convention on Maritime Search and Rescue (SAR Convention) 225*
11.2 *Some Significant Changes to Support MASS Operations 225*
    11.2.1 *International Code for Ships Operating in Polar Waters (Polar Code) 226*
    11.2.2 *International Maritime Dangerous Goods (IMDG Code) 226*
    11.2.3 *International Bulk Chemical (IBC Code) 226*
    11.2.4 *International Code for the Construction and Equipment of Ships Carrying Liquefied Gasses in Bulk (IGC Code) 227*

11.2.5 International Code on the Enhanced Programme of Inspections During Surveys of Bulk Carriers of Oil Tankers (ESP Code) 228
11.2.6 International Code for the Safe Carriage of Packaged Irradiated Nuclear Fuel, Plutonium and High-Level Radioactive Wastes on board Ships (INF Code) 228
11.2.7 International Code for Fire Safety Systems (FSS Code) 228
11.2.8 Code of the International Standards and Recommended Practices for a Safety Investigation into a Casualty or Marine Incident (Casualty Investigation Code) 229
11.2.9 International Maritime Solid Bulk Cargoes (IMSBC) Code 230
11.2.10 International Code for the Safe Carriage of Grain in Bulk (International Grain Code) 230
11.2.11 Code of Safe Practice for Cargo Stowage and Securing (CSS) Code 231
11.2.12 International Convention on Standards of Training, Certification and Watchkeeping for Fishing Vessel Personnel (STCW-F Convention) 231
11.2.13 International Convention on Load Lines (LL Convention) 232
11.3 Little or No Significance to Regulations to Support MASS Operations 232
11.3.1 IMO Instruments Implementation (III Code) 233
11.3.2 International Code on Intact Stability (IS Code) 233
11.3.3 International Code for Application of Fire Test Procedures (FTP Code) 233
11.3.4 International Life-Saving Appliance (LSA Code) 233
11.3.5 Code for Recognized Organizations (RO Code) 234
11.3.6 International Convention on Tonnage Measurement of Ships (Tonnage Convention) 234
11.3.7 International Convention for Safe Containers (CSC Convention) 234
11.4 Instruments Not Yet Considered 235
11.4.1 International Convention for the Prevention of Pollution from Ships (MARPOL Convention) 235
11.4.2 International Code of Safety for High-Speed Craft (HSC Code) 235

　　　　11.4.3 *International Code of Safety for Ships Using Gasses or Other Low-Flashpoint Fuels (IGF Code) 235*
　　　　11.4.4 *Code for the Construction and Equipment of Ships Carrying Dangerous Chemicals in Bulk (BCH Code) 235*
　　　　11.4.5 *Special Trade Passenger Ships Agreement (STP Agreement) 236*
　　　　11.4.6 *Protocol on Space Requirements for Special Trade Passenger Ships (Space STP Protocol) 236*
　　　　11.4.7 *Code of Safety for Nuclear Merchant Ships 236*
　　*References 236*

**12 Future Directions of MASS**　　　　　　　　　　　　　　　**239**
　　12.1　*Demonstrated Competency of MASS 240*
　　12.2　*Fitness for Duty 241*
　　12.3　*Security 241*
　　12.4　*Environmental Concerns 243*
　　　　12.4.1 *5G Broadband Technology 243*
　　　　12.4.2 *Contribution of Greenhouse Gas to the Environment 244*
　　12.5　*Smart Ports 245*
　　12.6　*Aids to Navigation 245*
　　12.7　*MASS Operator Complacency 246*
　　12.8　*Is IMO Degree 4 Full Automation for MASS Ethical, or Even Possible? 248*
　　12.9　*Situational Awareness below the Waterline 248*
　　12.10 *Crowdsourcing MASS Subsea Sensor Data 250*
　　12.11 *MASS Will Lead Shipping into the Future 250*
　　12.12 *Post-IMO Regulatory Scoping Exercise 251*
　　*References 252*

*Index*　　　　　　　　　　　　　　　　　　　　　　　　　　　**255**

# Preface

Unmanned and autonomous ships are quickly becoming a reality in an effort to make shipping safer and more efficient. However, traditional lines based upon function and scale become blurred as new technology changes how the unique needs of different sectors are met. In addition to large vessels dedicated to the transport of goods and cargos across the oceans, major efforts are underway towards automation of small coastal shipping that includes ferries, tugboats, supply and service vessels, and barges. Further innovation is taking place in automated vessels replacing conventional ships for inspecting and servicing pipelines, drilling platforms, wind farms and other offshore installations. Tasks that have in the past been performed by large ships are planned to be accomplished by much smaller vessels, surface and undersea vehicles designed for specific purposes. Innovations in hydrodynamics and aerodynamics are leading to the development of unmanned automated hybrid vessels that appear to be a cross between ships and airplanes, sharing the advantages and disadvantages of both.

The segment of unmanned and automated ships within the responsibilities of both nation states and the International Maritime Organization (IMO) is considered in this book and, specifically, Maritime Autonomous Surface Ship (MASS) operations that may be addressed by existing and future IMO instruments as adopted and implemented in national regulations. The subject is introduced in Chapter 1 by providing a historical perspective on shipping and framing current initiatives in terms of ship, technology and test bed development within this context. In Chapter 2 automated shipping is explored in terms of economics, technology, safety and the environment. Chapters 3 through 8 discuss details of autonomy and automation, ship design and engineering, command and control, navigation, communications, training and security as pertain specifically to this unique segment of the shipping industry. Chapter 10 is dedicated to describing regulatory issues at the international level, among nation states, classification societies and non-governmental organizations and in Chapter 11 there is a discussion of the legal instruments and issues that have been considered with regard to automated maritime operations. Finally, in Chapter 12 future short- and long-term directions for unmanned and autonomous shipping are described, including predictions for civilian adoption of military ship automation programs, new trends in instrumentation and recommendations as to how mariners, industry and governments can enhance the opportunities presented through the adoption of these technologies.

This book is aimed at mariners, ship owners and operators, regulatory authorities, protection and indemnity insurance clubs, environmental groups and others interested in maritime affairs. Others with interest include undergraduate students involved in deck officer training, graduate students and academics involved in research pertaining to ship design, navigation and environmental studies.

# Acronyms and Abbreviations

| | |
|---|---|
| ABS | American Bureau of Shipping |
| ACTUV | Anti-submarine warfare Continuous Trail Unmanned Vessel |
| AMOS | NTNU Center for Autonomous Operations and Services |
| AI | Artificial Intelligence |
| AIS | Automated Identification System |
| AIS-ATON | AIS (radio-based) ATON |
| ARPA | Automatic Radar Plotting Aid |
| ASDS | Autonomous Spaceport Drone Ship |
| ATON | Aid to Navigation |
| BeiDou | Navigation Satellite System (China) |
| BIMCO | Baltic and International Maritime Council |
| BV | Bureau Veritas |
| CAD | Computer Aided Design |
| CAE | Computer Aided Engineering |
| CAM | Computer Aided Manufacturing |
| CCS | China Classification Society |
| Cefor | Nordic Association of Marine Insurers |
| CMI | Comité Maritime International |
| COLREGS | International Regulations for Preventing Collisions at Sea |
| CPA | Closest Point of Approach |
| CPU | Central Processing Unit |
| CSBWG | Crowd Sourced Bathymetry Working Group (International Hydrographic Organization) |
| CSO | Company Security Officer |
| DARPA | Defense Advanced Research Projects Agency (United States) |
| DMA | Danish Maritime Authority |
| DNV-GL | Det Norske Veritas/Germanischer Lloyd |
| DSC | Digital Select Calling |
| ECDIS | Electronic Chart Display Information System |
| eLoran | Enhanced Long Range Navigation system |
| ENC | Electronic Navigational Chart |
| FFI | Norwegian Defence Research Establishment |
| Galileo | Satellite Navigation System (European Union) |

| | |
|---|---|
| Gbps | Gigabits per second |
| GHz | Gigahertz |
| GLONASS | Globalnaya Navigazionnaya Sputnikovaya Sistema (Russia) |
| GMDSS | Global Maritime Distress and Safety System |
| GNSS | Global Navigation Satellite System GPS Global Positioning System (United States) |
| GPU | Graphical Processing Unit |
| IACS | International Association of Classification Societies Ltd |
| IALA | International Association of Marine Aids to Navigation and Lighthouse Authorities |
| IBS | Integrated Bridge System |
| ICS | International Chamber of Shipping |
| IFSMA | International Federation of Shipmaster's Associations |
| IMarEST | Institute of Marine Engineering, Science and Technology |
| IMCA | International Marine Contracting Association |
| IMO | International Maritime Organization |
| INS | Integrated Navigation System |
| IoT | Internet of Things |
| ISO | International Organization for Standardization |
| ISMFA | International Federation of Shipmasters' Associations |
| ISO | International Organization for Standardization |
| ITF | International Transport Workers' Federation kHz kilohertz |
| KR | Korean Register of Shipping |
| LEO | Low Earth Orbit |
| Lidar | Light Imaging Detection and Ranging |
| LNG | Liquefied Natural Gas |
| Loran | Long Range Navigation system |
| LL | International Convention of Load Lines |
| LR | Lloyd's Register of Shipping |
| LRIT | Long Range Identification and Tracking |
| MARINTEK | Norwegian Marine Technology Research Institute |
| MARPOL | International Convention for the Prevention of Pollution from Ships |
| MARSEC | Maritime Security |
| MASS | Maritime Autonomous Surface Ships |
| Mbps | Megabits per second |
| METOC | Meteorological and Oceanographic |
| MHz | Megahertz |
| MSC | Maritime Safety Committee (IMO) |
| MSI | Maritime Safety Information |
| MUNIN | Maritime Unmanned Navigation through Intelligence in Networks |
| NCA | Coastal Administration |

| | |
|---|---|
| NGO | Non-Governmental Organization |
| NFAS | Norwegian Forum for Autonomous Ships |
| NI | Nautical Institute |
| NMA | Norwegian Maritime Authority |
| NOAA | National Oceanographic and Atmospheric Administration (United States) |
| NPU | Neural Processing Unit |
| NTNU | Norwegian University of Science and Technology |
| OPU | Optimizing Processing Unit |
| OOW | Officer of the Watch |
| PXI | PCI eXtensions for Instrumentation |
| RACON | Radar Beacon |
| Radar | Radio Detection and Ranging |
| RCC | Remote Control Center |
| SAE | Society of Automotive Engineers |
| SAR | Search and Rescue |
| SOLAS | International Convention for the Safety of Life at Sea |
| Sonar | Sound Navigation and Ranging |
| SSC | Smart Ships Coalition |
| STCW | International Convention on Standards of Training, Certification and Watchkeeping |
| UAV | Unmanned Aerial Vehicle |
| UCSDA | Unmanned Cargo Ship Development Alliance |
| USV | Unmanned Surface Vehicle |
| UUV | Unmanned Underwater Vehicle |
| VATON | Virtual Aid to Navigation |
| VSO | Vessel Security Officer |
| VSP | Vessel Security Plan |
| VTS | Vessel Traffic Services |
| WAN | Wide Area Network |

# Chapter 1
# Introduction

This book examines unmanned and autonomous shipping within the context of key design elements and systems necessary to accomplish multiple levels of automation. We begin with this chapter by providing relevant history and cite examples of present-day developments along with some topics of concern voiced by mariners. Subsequent chapters examine the future of maritime shipping and its impact on trade, national economies, technological advancement and the seafarers who have bravely pursued their trades over the centuries. Automation is being accomplished across all levels of shipping from the smallest recreational boats to the largest oil tankers, container ships, cargo carriers and cruise ships. While the contents of this book generally apply to vessels of all sizes and types supporting diverse functions and purposes, its goal is to examine the issues associated with larger surface vessels and their support craft that are operated by professional captains, mates, crews and pilots. No attempt is made to address issues associated with small drones, more commonly referred to as unmanned or autonomous surface vehicles (USVs/ASVs), except as they may be utilized by large vessels, even though there may be great commonality in the technical details of their implementation.

A question exists of whether unmanned and autonomous ships represent the next step in the logical evolution of shipping and ship technology or if they exemplify disruptive innovation that will completely transform the face of maritime shipping as we know it. Although there are vested interests promoting both scenarios, rapid change generally does not occur in the shipping industry and there is no reason to believe that present-day masters and crews are destined for early retirement. Maritime jobs continue to be influenced by changes in technology and economies and few would or even should take the time to argue over the virtues of oar versus sail, steam versus gas turbine, or nuclear versus liquefied natural gas (LNG) or hydrogen fuel cell propulsion. Some of these technologies are relegated to the past while others represent possible alternatives for the future. In the meantime, it is hoped this book can draw attention to the present state of events and activities surrounding the furtherance of what the International Maritime Organization (IMO) refers to as Maritime Autonomous Surface Ships

(MASS). Readers may then decide for themselves as to what the future may hold and what influence they have had.

## 1.1 A HISTORICAL PERSPECTIVE ON ADVANCES IN SHIPPING

There are certain events considered key milestones in the development of ships and shipping technology that were hallmarks of change. Sails harnessed the wind to propel ships across vast distances, while oars provided a means of propulsion that made vessels more maneuverable and could be depended upon when the wind was unreliable. Iron ships eventually replaced wooden ships, and steam power replaced sails in general but wind-powered ships using new technologies are still being actively developed. These changes occurred gradually with iron being used initially to connect fasteners and strengthen hulls decades before the appearance of the first iron ships. However, new technology also introduced new problems. One example is how iron ships adversely affected the operation of magnetic compasses. Through the experimentation of astronomer Sir George Biddell Airy in 1839, this problem was solved with his development of correction tables that were subsequently adopted by the Merchant and Royal Navies in England and spread throughout the rest of the world [Airy 1839; Wright 1988]. Such innovations originated from a broad cross-section of people and professions including naval architects as well as inventors, practical tradesmen and scientists, many of whom were not mariners, shipbuilders, or even seafarers. Nevertheless, their discoveries and inventions changed the face of shipping, opened up worldwide trade routes and set the stage for exploration and mass migration on a large scale. Today, similar advances are being made by engineers, metallurgists, chemists, computer scientists and others that may have never set foot on the deck of a ship. Yet, their contributions in the development of new electronic control and guidance systems along with the software through which they operate are essential for today's ships. The introduction of new technology to aid in vessel communication, navigation and overall situational awareness is summarized by the timeline shown in Figure 1.1. The timeline is not to scale and many of the dates given are approximations as conflicting dates and claims are often present in the literature. However, it provides a good illustration of how developments have accelerated during recent times in four primary disciplines: Construction, propulsion, navigation and communications. The convergence of many aspects of these four disciplines in the modern era has ultimately lead to the present day where the concept of MASS can now be entertained through the establishment of automated command and control over all of the systems needed to operate the vessel and therefore the entire vessel itself.

The following paragraphs provide brief descriptions of new developments in technology and ship function that have taken place relatively

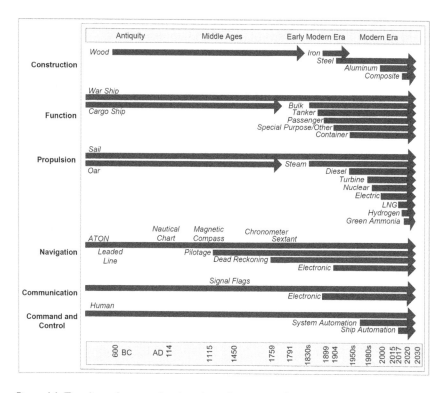

*Figure 1.1* Timeline of new technology introduction on ships.

independently of each other, yet are essential components to achieve vessel autonomy. The common denominator in modern times among these disciplines is the advent of computer-aided design and simulation of all aspects of ship construction and performance combined with electrical and electronic control of all major ship propulsion, navigation and communications systems.

### 1.1.1 Construction

Wood floats, and it was a natural progression to build log canoes to large sailing vessels from this material. Towards the late 1700s iron plates began to appear on inland barges, in part out of necessity due to shortages of lumber [Walker 2010]. Ships with hulls made of iron started to appear in 1818, when the iron barge *Vulcan* was built near Glasgow, Scotland. Overcoming prejudices and concerns among early seafarers as to whether iron ships could actually float was a key inhibitor of early progress [Benson 1923]. Once new processes were developed for manufacturing steel and it became available in sufficient quantities (i.e., tons), large plates of various thicknesses were used to build ships of steel. Improvements in riveting and welding helped to increase vessel size from the 5,000-ton ships of the 1880s to the 250,000-plus-ton ships that are commonplace today. In some cases, attempts have been successful in creating hulls using aluminum. However, difficulties related to welding and galvanic corrosion, especially where a steel shaft passes through the hull, reduce reliability and impose greater maintenance requirements. Carbon fiber composite materials providing lightweight and strong hull forms are presently being developed for smaller vessels but its use for large ships is yet to be realized due to the high costs involved in its production.

Of greatest significance to make possible automated ships are the revolutionary changes in modern era methods and processes used in their construction. As ships have increased in size the former methods of using curves, drawing frames and templates have given way to computer-aided design (CAD), computer-aided engineering (CAE) and computer-aided manufacturing (CAM), making it possible to visualize, evaluate and test design concepts long before the first metal is cut. This includes determining the types of onboard physical and virtual sensors necessary and their proper placement to assess ship performance. The outcome includes better processes that result in improved design and production times at less cost than would otherwise be possible.

### 1.1.2 Function

Ship designs originated to perform two distinct functions as warships and cargo ships, each with their own unique design features that made them suitable for their task. Cargo ships were also used to transport small numbers of passengers. As the use of sails gave way to steam propulsion, ships of different designs began to be developed for the transportation of inland freight and passengers. Cargo ships continued to develop to support more specialized cargos separating into bulk carriers, tankers and container ships in the 20th century. During this same period passenger ships, with and without cargo, also began to flourish from small river boats and passenger ferries, to great ocean liners and cruise ships designed for pleasure excursions. In the meantime, other specialized ships for fishing, offshore construction and support, drilling, cable laying and a myriad of other uses entered the scene. A broad cross-section of vessels performing these functions is the focus of many autonomy research and development efforts today.

### 1.1.3 Propulsion

Sails have been present on ships since antiquity and are historically the primary mode of ship propulsion up until modern times. Wind power provided by rotors and other inventions is still being considered for auxiliary ship propulsion. The use of paddles and oars has also complemented sail power. However, the advent of the steam engine heralded the modern age of ship propulsion. Early problems with steam engines and their lack of power were eventually overcome as experience and lessons learned in their use were acquired. However, other problems persisted. Robert Fulton's steamboat *Clermont*, otherwise known as Fulton's Folly, caused spectators to fear when the smoke, flames and steam of the engine made it appear about to explode [Sale 2017]. The 1900s found coal-fired boilers replaced with oil, and today steam has become obsolete and replaced by diesel and turbine power. Nuclear power was used on an experimental basis in the mid-1900s to power commercial ships. However, except for a few specialized ships such as icebreakers, its maritime use is generally relegated to warships. Today on many vessels large engines connected to propellers using drive shafts are being replaced with electric engines connected directly to a propeller enclosed in a steerable gondola or pod suspended below the hull. Liquefied natural gas (LNG) is appearing as a fuel of choice that provides much cleaner performance in terms of environmental emissions. Hydrogen and green ammonia are other potential clean fuel sources being considered for the future.

Just as significant as the types of engines and fuels used on board ships has been the evolution of automated engine and power monitoring and control systems that enable many ships to operate with reduced staffing requirements and unattended machinery spaces. These electronic systems can automatically monitor and control temperatures, pressures, flows, levels, torque and other characteristics of propulsion systems necessary for safe operation.

### 1.1.4 Navigation

Basic navigation capabilities have been achieved through the invention of the leaded line, nautical chart, compass, chronometer and sextant. The leaded line used to measure water depth was mentioned by Herodotus in the fifth century B.C. [Macaulay 1890]. Early marine charts can be traced back to Marinus of Tyre in the second century [Deetz 1943]. The compass was used for maritime navigation in China around the year 1115 [Ronan 1986]. The chronometer, used for obtaining precise timing information, became available in 1759 [Gould 1921]. The modern sextant had its origins around 1791 [Ifland 1998]. These inventions form the core instruments found in all modern ships. However, the use of electronic systems in the modern era has changed the face of ship navigation.

### 1.1.5 Communications

The primary method of communication between ships began in the 15th century with the appearance of systems of signal flags and pennants hoisted for communication [Sterling 2011]. These were eventually replaced in the

19th century with the advent of semaphore systems. Light signals were used at night and became more viable with the invention of the electric light bulb. Paralleling advances in navigation, communications by ships with land sites and between ships greatly improved with the invention of the wireless telegraph and other electrical and electronic systems.

### 1.1.6 Electronic Navigation and Communications

Although slow to occur at first, the pace of evolution of navigation and communications technology has dramatically accelerated beginning in the last century with the advent of electrical and electronic devices as illustrated in Table 1.1. The introduction of radio detection and ranging (Radar) equipment for shipboard use occurred in 1937 [Eagle 2008]. The effective use of Radar on ships was stymied by a lack of standards and training that eventually resulted in the first Radar-assisted collision between *Andrea Doria* and *Stockholm* [Meurn 2013]. Techniques were developed over the decades to plot range and bearings to stationary landmarks and buoys to establish position fixes for navigation, and to other ships as a means of collision avoidance. Radar beacon (RACON) transponders that reside on buoys and structures and emit a signal when interrogated by a Radar transmission were introduced in 1962 [ITU-R M.824-4]. In 1979, Automatic Radar Plotting Aid (ARPA) technology was introduced to acquire and track targets and automatically compute and display information to aid the bridge watch in collision risk assessment and collision avoidance [Hayashi et al. 1994].

Automated Identification System (AIS) was introduced in 2000 to automatically exchange static and dynamic ship information pertaining to the voyage, safety and security between vessels and to shore stations [Noris 2008]. AIS has also been introduced onto the Radar display whereby ship name, speed, heading and other information may be shown, aiding in target identification and providing a means to establish direct communications with specific vessels of interest to exchange passing and other information [Pillich and Schack 2002]. This was followed by the introduction of satellite AIS beginning in 2005 for the tracking of vessels at sea beyond the range of shore-based receivers [ESA 2009].

Electronic Chart Display Information System (ECDIS) is a navigation information system intended to display all chart information necessary for safe and efficient navigation originated by, and distributed on the authority of, government-authorized hydrographic offices [MSC 82/24]. With adequate back-up arrangements it may be accepted as complying with the Convention on the Safety of Life at Sea (SOLAS) up-to-date chart requirements [SOLAS 1974]. ECDIS provides access to the electronic equivalent of a paper chart for vessel navigation through the means of an electronic navigation chart (ENC) and system electronic navigation chart (SENC) comprised of database attributes used for display generation and navigation

Introduction 7

Table 1.1 Innovations in Electronic Communication and Navigation

| Communication | | | Navigation | | | | |
|---|---|---|---|---|---|---|---|
| Invention | | First Use | Invention | | First Use | Invention | First Use |
| Radio and Satellite | Wireless Telegraph | 1899 | Terrestrial | Echosounder | 1904 | RADAR RADAR | 1937 |
| | MF-HF Radio | 1907 | | Radio Time Standards | 1905 | RAMARK / RACON | 1962 |
| | VHF Radio | 1940 | | Radio Direction Finder | 1913 | ARPA | 1979 |
| | SATCOM | 1979 | | Radiobeacons | 1917 | Satellite NAVSAT / TRANSIT | 1960 |
| | GMDSS | 1988 | | LORAN | 1943 | GPS | 1979 |
| | NAVTEX | 1993 | | DECCA | 1943 | Differential GPS | 1990 |
| | AIS | 2000 | | Inertial Navigaation | 1952 | GLONASS | 2012 |
| | Satellite AIS | 2005 | | OMEGA | 1971 | Galileo | 2012 |
| | Maritime Cloud | 2017 | | eLORAN | 1997 | BeiDou | 2015 |
| | | | | Navigation (FL) Sonar | 1997 | ECDIS | 1995 |
| | | | | AIS-Aids to Navigation | 2012 | eNavigation | 2005 |

functions. However, ECDIS is more than a device used to display chart information. It is also capable of displaying Radar and ARPA, AIS, Global Navigation Satellite System (GNSS) positioning, speed log, heading, depth and other navigational information. Final implementation of ECDIS vessel installation and training requirements occurred in January 2017 [STCW 2010].

Many navigation and communication systems are interconnected through a standard data bus like the National Electronics Manufacturers Association (NEMA) 2000 plug-and-play standard that communicate data from marine sensors to computers and displays. These and future standards that support greater bandwidth for the transmission of imagery and other sensor information will form the backbone for MASS automation by which all details of vessel systems and sensors may be communicated and controlled.

## 1.1.7 Command and Control

Over the last 20 years great innovation has taken place in the automation of ship controls and systems in terms of engine operation, power management, navigation and communications. Computer-aided design and simulation of all aspects of ship construction has ensured that adequate provisions are made for the installation, powering and transfer of information among individual system components as well as between seemingly disparate systems that have seemingly little or no relationship in form or function. Engine room systems generally operate in relative harmony without constant human direction as the navigation systems follow a predetermined course to arrive at their destination with seemingly very little effort. Communications is routinely accomplished whereby enormous amounts of data are automatically transferred internally as well as to and from the ship.

In addition to providing more optimal processes resulting in improved design and production times at less cost than would otherwise be possible, CAD/CAE computer models can be extended to encompass all systems throughout the ship, making possible the fusion of propulsion, power distribution, navigation, communications and other functions. Intelligent reasoning systems can evaluate the data and information acquired from these systems and the sensors upon which they rely to draw inferences and conclusions from which decisions may be made. At present this decision-making process is accomplished by the ship's Captain in consultation with mates and crew based upon their specialized skills and knowledge. His or her comprehensive knowledge of the ship itself and its capabilities forms the basis for all command reasoning and decision making, considering the characteristics of the course to be followed, the rules of the road, weather conditions and any possible threats or hazards that may be encountered while in transit. The goal of MASS is to realize this same decision-making process at increasing levels of autonomy in an effort to increase safety of navigation, reduce costs and gain other efficiencies. Furthermore, lessons learned and tools developed in achieving autonomy can be applied across all modes of conventional shipping and ship handling. The overall results should benefit the shipping industry and seafarers as a whole.

## 1.2 CURRENT INITIATIVES IN UNMANNED AND AUTONOMOUS SHIPPING

The present state of MASS involves broad initiatives as well as very specific programs on the part of many worldwide organizations. Both industry and academia are currently involved in wide-ranging research and development (R&D) efforts to determine the technical feasibility and limits of technologies involved in MASS operation. Many governmental organizations support these R&D efforts and also participate in analyzing the scope of the current regulatory environment through the IMO to determine what existing regulations apply to MASS and what new regulations may be needed. Class societies are examining the needs to establish and maintain technical standards for the construction and operation of MASS, while many non-governmental organizations are active representing mariners, shipbuilders and operators, engineers and technologists, and other individuals and groups across a broad range of technical, environmental, legal and other interests. A summary of some of these initiatives is included here, with further details provided in later chapters and especially in Chapters 10 and 11.

### 1.2.1 Industry and Academia

There is at present a great deal of activity on the part of industry and academic organizations in R&D and the introduction of production autonomous ships and ship technology. Several such initiatives are cited below:

- Norway's Kongsberg development of the fully electric and autonomous container ship *Yara Birkeland*, illustrated in Figure 1.2a, for shortsea shipping [Hand 2017].
- U.S. shipbuilder Metal Shark partnership with autonomous vessel technology developer ASV Global to build their Sharktech autonomous vessels, illustrated in Figure 1.2b, in aluminum, steel and composite materials from 5 to almost 100 meters in length [Wingrove 2018]. They can be used for naval, surveillance, firefighting or survey in the military, law enforcement, fire rescue and commercial markets under multiple modes of autonomy including unmanned, reduced manned or conventional manned operations.
- Another project by the UK's ASV Global integrates their autonomous vessel control simulator and BMT Ship & Coastal Dynamics REMBRANDT ship maneuvering simulator into a single suite to address the challenge of how traditional manned vessels can co-exist with autonomous systems in shared water space [MAREX 2017a].

*Figure 1.2* Examples of MASS research and development projects (Photos with permission). (a) Kongsberg Autonomous Vessel *Yara Birkeland*. (b) Sharktech Autonomous Vessel. (c) ASL Autonomous Vessel *Hrönn*. (d) Robert Allan Ltd (RAL) *RAmora* tug.

- The offshore service company Bourbon and Automated Ships Ltd partnered together to support the building of *Hrönn*, illustrated in Figure 1.2c. This was planned to be the world's first autonomous, fully automated and cost-efficient prototype vessel for offshore operations, developed in collaboration with the project's primary technology partner, Kongsberg [MAREX 2017b].
- Pacific Maritime Institute and Robert Allan Ltd have completed comprehensive testing of control systems for the remotely operated *RAmora* tug, illustrated in Figure 1.2d, for ship-assist and berthing operations in high-risk situations [MAREX 2016].
- The European Commission funded the Maritime Unmanned Navigation through Intelligence in Networks (MUNIN) project as a joint collaboration between Chalmers Technical University (Sweden), University College Cork (Ireland) and several industrial organizations to verify a concept for an autonomous ship [MUNIN 2016].
- Rolls-Royce partnered with Google to develop Rolls-Royce's intelligent awareness software that will comprise a key role in the company's drive towards autonomous vessels [MAREX 2017c]. Under their agreement Rolls-Royce would use the Google Cloud Machine Learning Engine to train Rolls-Royce's artificial intelligence (AI)-based object classification system to detect, identify and track surface objects.
- Sea Machines of Boston Massachusetts in the United States is the developer of autonomous technology that specializes in advanced control technology for workboats and other surface vessels [Sea Machines 2019]. In July 2019, they entered into a cooperative agreement with the U.S. Maritime Administration to test the use of autonomous technology in response to marine oil-spills [gCaptain 2019].
- The *Mayflower* autonomous ship being developed by Promare in Chester, Connecticut, in the United States is a 30-meter hybrid powered vessel being built in Poland that will sail between Plymouth England and Plymouth Massachusetts, United States, in 2020 [Promare 2019].
- Industry and academia have also participated in the development and successful demonstration of the 40-meter-long *Sea Hunter*, which is one of the most advanced autonomous unmanned ship projects under the U.S. Office of Naval Research [ACTUV 2018]. Created as part of the Anti-Submarine Warfare (ASW) Continuous Trail Unmanned Vessel (ACTUV) program, this represents an entirely new class of ocean-going vessel able to traverse thousands of miles over open seas for months at a time, without a single crew member aboard. In February 2019, *Sea Hunter* sailed fully autonomously from California to Hawaii and back [HNN 2019].
- The U.S. Navy is also developing Medium Unmanned Surface Vehicles (MUSV) of between 12 to 50 meters in length, and Large Unmanned Surface Vehicles (LUSV) of greater than 50 meters beginning in 2019 and continuing through the 2020's [CRS 2019].

These are just a few examples of the pioneering work being accomplished worldwide in the development of MASS and in the advancement of MASS technology. From unmanned support craft, tugs and ferries to large vessels the results of research are being integrated across many types of ships to accomplish functions and tasks previously accomplished by seafarers. In some cases such as firefighting and law enforcement, these vessels can remain on station indefinitely and under conditions considered too hazardous for human presence. In shortsea shipping, MASS can theoretically perform routine shipments along well-defined transit corridors at lower cost than conventional shipping. In all cases the potential to eliminate human error is a powerful incentive to proceed with MASS. However, the exchange of human error for potentially unknown opportunities for machine error must be adequately addressed before MASS should proceed beyond the experimental stage without direct human supervision. However, as we have seen in the crashes of Lion Air Flight 610 and Ethiopian Airlines Flight 302 in 2018 and 2019 of the Boeing 737 MAX 8, human attempts at overriding "safety" features by performing emergency procedures may not always prove successful in systems that have not been properly verified as being fit for use and service [Patterson 2019].

### 1.2.2 Regulatory Authorities

The IMO is leading efforts to establish regulations that pertain to the unique requirements of MASS through the initiation of a regulatory scoping exercise intended to determine what regulations appear to already apply and to identify areas in which new regulations may be needed. This exercise began at the 97th session of the Maritime Safety Committee's (MSC) where a new agenda item was established to amend the regulatory framework to enable the safe, secure and environmental operation of partly or entirely unmanned MASS [MSC 98/20/2]. This included their interaction and co-existence with manned ships within the existing framework. This exercise was completed to a great extent at the 101st session of the MSC in June 2019 followed by the meeting of the MSC Intersessional Working Group on MASS in September 2019. Contributing and participating at these meetings were the European Union represented by Denmark, Finland, Norway, Sweden and the United Kingdom, as well as China, Singapore, Japan and the United States. Also participating were Australia, Canada, Estonia, France, Liberia and South Africa. Upon concluding the regulatory scoping exercise at MSC-102 in May 2020, the hard work began to adapt existing and create new rules and conventions for MASS.

### 1.2.3 Classification Societies

Classification societies contribute to MASS discussions both as individual organizations and as part of the Unmanned Cargo Ship Development Alliance (UCSDA), which is also made up of shipyards, equipment manufacturers and designers. Participating in the IMO regulatory scoping exercise were the International Association of Classification Societies (IACS), American Bureau of Shipping (ABS), Bureau Veritas (BV), China Classification Society (CCS), Det Norske Veritas Germanischer Lloyd (DNV-GL), Korean Register of Shipping (KR) and Lloyds Register (LR). Several of these organizations are also reviewing their standards and services in accordance with developing MASS requirements.

### 1.2.4 Non-Governmental Organizations

Non-governmental organizations active in the development and promotion of MASS and their supporting technologies include the Baltic and International Maritime Council (BIMCO), Comité Maritime International (CMI), International Federation of Shipmasters' Associations (IFSMA), International Group of Protection and Indemnity Clubs (IGP&I), International Transport Workers' Federation (ITF), International Chamber of Shipping (ICS), Institute of Marine Engineering, Science and Technology (IMarEST), International Marine Contractors Association (IMCA), International Organization for Standardization (ISO), the Nautical Institute (NI), One Sea Autonomous Maritime Ecosystem, and the Smart Ships Coalition (SSC). These organizations represent a diverse community of seafarers, business, engineering and technical, standards, insurance and other professional interests.

## 1.3 A MARINER'S PERSPECTIVE

The maritime industry has undergone major changes throughout its history with the development of new hull forms optimized for trade and war, methods of propulsion, and new ways to navigate and to communicate between vessels and land sites. The next major change looming on the horizon is the digitalization and automation of ships and the functions they perform that heretofore have been performed by seafarers. On the one hand, the promises of automation have been dangled in front of workers for decades in terms of shorter working hours, more pay, more time for hobbies, etc. [Bellamy 1888]. One description of the potential benefits of properly implementing ship automation was expressed in terms of "driving a Mercedes, seeing wife & family each evening, ... is not that bad" rather than enduring long periods at sea [Bertram 2013]. Others are concerned about the elimination of traditional jobs and condemning mass scores of gainfully employed individuals to poverty [Smith and Anderson 2014]. From one perspective the evolution of shipping is proceeding along a well-explored path that is similar to other industries. Another perspective cites the unique needs and requirements of shipping that make it completely different from other industries. Like all extreme views, the truth is very likely to exist somewhere in the middle. The following paragraphs attempt to address some of these important issues.

### 1.3.1 Human Senses Exceed Remote Operator and Full Autonomy Capabilities

There exists much concern and anxiety regarding MASS and the ability to implement partial or fully automated vessels without having human presence on board. The implausibility of a remote operator limited to displays of information provided by sensors and a communication link of attaining the same level of situational awareness and safety as an onboard operator monitoring the same displays was the subject raised by the IFSMA and ITF in their contribution to the discussion on MASS at the IMO [MSC 99/5/1]. The premise of the argument is that the onboard operator has the additional ability to validate the displayed information against actual observation of the real-world environment and feel of the ship with the ability to physically intervene.

This concern correctly describes the immeasurable qualities of human beings to sense a wide range of conditions that are just slightly out of harmony or balance in terms of the feel of a ship's movements through the water, the vibrations of onboard machinery, the whistling of wind through the rigging, and even the sight of low dark clouds along the horizon as a portent of bad weather that was not previously forecast. The response to this question raises another question of whether "the same level of situational awareness and safety" onboard today's manned vessels is in itself adequate. Based upon the nature, types and frequency of maritime accidents that continue to occur due to human error it would be presumptuous to assume watchstander situational awareness is anything but lacking. This is not to necessarily blame the watchstanders for this situation as their basic knowledge and skills have been honed over centuries of experience imparted unto them through rigorous training curriculums developed in consultation with a wide range of worldwide maritime authorities. Rather, it would be beneficial to consider the scope of their duties in terms of the sprawling physical environment within which they must operate as well as the complexity and volume of the stimulus and alarms they must respond to on a continuous basis. Perhaps it is time to revaluate exactly how technology may aid in expanding situational awareness using the full capabilities of modern sensors not only for MASS, but also to assist watchstanders on manned ships today and in the future.

One imperative of ship automation is the need to equip such vessels with networks of smart sensors far beyond those currently mandated that can analyze data and draw conclusions regarding phenomenon occurring in real time, and notify the remote operator and onboard reasoning system about the existence and nature of anomalies detected. This includes sensor systems capable of expanding sight throughout and adjacent to the ship across a broad range of the visible, infrared and radio frequency spectrum to detect movement and identify specific people, events and situations under both normal and abnormal conditions. Also needed are vibration monitoring, fire detection, sound monitoring and other topside sensors as well as on the hull both above and below the waterline to monitor many aspects of physical phenomena that exist on vessels of all types. One critical example includes expanding situational awareness below the waterline ahead in the path of transit using forward-looking navigation sonar to detect, identify and avoid hazards to navigation such as shoals, reefs, shipping containers adrift, growlers and icebergs, and even large marine mammals. Sonar data can also be archived and mined to determine changes that occur over time that may be indicative of natural hazards and wrecks not detected since the last hydrographic survey or listed in a navigation chart. Indeed, high-resolution swath bathymetry acquired from navigation sonar should be shared through crowdsourcing initiatives such as the International Hydrographic Organization (IHO) Crowdsourced Bathymetry Working Group to help improve navigation charts [CSBWG 2018].

The fusion of various smart sensor data and the supplying of verifiable and actionable conclusions to a remote operator or onboard reasoning system far surpass the limited information presented on a display to a bridge watchstander or remote operator. The development of such systems will not only benefit autonomous ships, but manned ships as well. However, this is not to say this process will be accomplished quickly. Indeed, the aviation industry is far more advanced in autonomous operations than is present-day shipping. Today's aircraft are fully capable of autonomous operation throughout their entire journeys as are modern spacecraft that return to earth unaided by human intervention. Yet these air and space vehicles are limited not due to technological issues but by human resistance to placing one's life in the hands of a computer – and rightly so! The same can be said for driverless car technology where unattended vehicles race through intersections, run into other vehicles and kill pedestrians crossing the street. The loss of hundreds of lives due to a computer error resulting in the crash of an airliner is no more acceptable than the grounding of a MASS that results in the loss of lives of search-and-rescue personnel, the property being transported, and the economic and ecological damage resulting from fouled shorelines and the death of marine wildlife.

### 1.3.2 Implicit Devaluation of the Maritime Professions

There is risk that the automation of many maritime professions may implicitly devalue the tasks performed and the achievements of those that presently perform their jobs. This concern spans all levels from deck crew members who by the look of a line can determine it is fraying in an abnormal way or that a wire rope is corroding and dangerous, to mates who notice intermittent glitches in the radar display that may result from unusual signal propagation conditions rather than a nearby target, to pilots whose direct knowledge of local marine conditions consistently helps to avoid groundings due to moving shoals, and captains whose vast years of experience have given them a sixth sense of how to avoid problems when dangerous conditions arise.

The question is not so much the future worth of intricate knowledge of present-day seafarers, but rather how to ensure this valuable knowledge may continue to be cultivated, captured and put to good use to create a corporate knowledge and capability base that is far superior to that of any one or group of individuals. An example can be cited in the automobile industry where automation has existed for decades. Instead of focusing on continually developing new and improved methods of how human technicians can perform innovative tasks, greater returns in terms of efficiency and economic value have been accomplished by exploring how human technicians guiding and aided by robotics applications can achieve greater innovation using the combined strengths of both. The day has not yet arrived where robotics combined with artificial intelligence (AI) will routinely result with the robots improving their own skills and performance, but that day is rapidly approaching. However, this will not be possible without the continuous availability and use of seafarer knowledge and experience.

The path to future ship automation is not likely to occur rapidly and there is great potential for expanding the value of seafarers into teaching and mentoring roles that can aid in the automation process – all while continuing to perform their existing jobs. Without the human expertise needed to perform tasks, explicitly through direction or implicitly by explaining the processes involved by example, there is little a machine can do to emulate these processes exclusive of exhaustive datasets counted in the millions of images and sensor measurements that are relatively unavailable or impossible to acquire. It is also unlikely that automated processes will readily acclimate to changes resulting from new situations and technologies that provide solutions to unanticipated problems. However, even these problems are being solved through new research results in AI demonstrating the ability to train themselves and to create human faces and other objects that are indistinguishable from the real versions [Ray 2019; Diaz 2018]. In the meantime, seafarers will continue to be needed and trained in their basic skill sets far into the future while they also acquire new skills necessary to adapt to new products and technologies needed to enhance their effectiveness in their jobs. This will include working with automated processes that should reduce their workload as well as improve the safety of their occupations.

### 1.3.3 Maritime Jobs

By all accounts the future outlook for maritime employment remains bright with the demand for experienced mariners exceeding the supply by comfortable margins. One report by the ICS projects a shortage of 147,500 maritime officers by 2025 based upon current average operational manning levels and other factors [ICS 2016]. These forecasts are not likely to change significantly due to the introduction of autonomous shipping. The world merchant fleet consisted of over 53,000 ships in 2018 and is expected to increase in proportion with improved world trade [Statistica 2019]. The prospects for the introduction of MASS exceeding more than one or two percent of the fleet (i.e., greater than 1,000 ships) over the next 20 years are minimal. A similar analogy may be made between mariners and truck and bus drivers with the introduction of automated vehicles. Even if automated vehicle trials are successful in the near term, experienced drivers will continue to be needed for many years into the future as the number of automated trucks and buses remains a small portion of overall fleet size. Moreover, ships are much more complex than automobiles and trucks and require far more than a driver for their safe operation.

Rather than automation drastically reducing the number of available maritime jobs, the advent of MASS should serve to increase maritime jobs by expanding opportunities into technical fields that will supplement the traditional maritime professions. Furthermore, new opportunities will be created for today's mariners to transition into roles that previously have not existed in the remote operation and monitoring of MASS similar to military aircraft pilots transitioning into drone pilots while many aircraft cockpits remain empty due to a shortage of pilots. Additional and new types of maritime jobs expected to be created due to the economic benefits MASS research, development and production are more fully discussed in Chapter 2.

### 1.3.4 Other Issues

The topics discussed in this part of the introduction serve as initial points of discussion for the rest of the book. Attempts have been made to identify issues most relevant to MASS and organize them into the individual chapters that follow where they may be covered in adequate detail.

**REFERENCES**

ACTUV 2018. ACTUV "Sea Hunter" Prototype Transitions to Office of Naval Research for Further Development. Defense Advanced Research Projects Agency, 30 January 2018. https://www.darpa.mil/news-events/2018-01-30a.
Airy, George Biddell. 1839. *Phil. Trans. Royal Society of London*, 24 April 1839.
Bellamy, Edward. 1888. *Looking Backward: 2000-1887*, Houghton-Mifflin, 1888. Republished version with a forward by Erich Fromm. Signer, 1960. ISBN 0-451-52412-8.
Benson, W. S., RAdm. 1923. The Merchant Marine. *Journal of the American-Irish Historical* Society, 22: 77.
Bertram, Volker. 2013. Towards Unmanned Ships. DNV-GL, 2013, p. 42. www.ntnu.edu/documents/20587845/1266707380/UnmannedShips.pdf.

CRS 2019. Navy Large Unmanned Surface and Undersea Vehicles: Background and Issues for Congress. Updated December 17, 2019. Congressional Research Service. https://crsreports.congress.gov. R45757.

CSBWG 2018. Crowd-Sourced Bathymetry Working Group (CSBWG). International Hydrographic Organization (IHO), Terms of Reference. 7 June 2018. www.iho.int/mtg_docs/com_wg/TOR/CSBWG_TOR.pdf.

Deetz 1943. Cartography 4, Special Publication No. 205, U.S. Coast and Geodetic Survey, 1943.

Diaz, Jesus. 2018. Nvidia's Scary AI Generates Humans That Look Real. Toms Guide, 17 December 2018. https://www.tomsguide.com/us/nvidia-ai-faces-generative-adversarial-network,news-28869.html.

Eagle 2008. Sunday Ship History: Putting Radar on Ships, 3 February 2008. http://www.eaglespeak.us/2008/01/sunday-ship-history-putting-radar-on.html.

ESA 2009. ESA Satellite Receiver Brings Worldwide Sea Traffic Tracking Within Reach. ESA, 23 April 2009.

*gCaptain* 2019. Sea Machines and MARAD to Demonstrate Autonomous Tech in Oil-Spill Response Operations. *gCaptain*, 24 July 2019. gcaptain.com/sea-machines-and-marad-to-demonstrate-autonomous-tech-in-oil-spill-response-operations/.

Gould, Rupert T. 1921. The History of the Chronometer. *The Geographical Journal*, 57, no. 4: 253–68.

Hand, Marcus. 2017. Autonomous Vessels a "Huge Opportunity" for Shortsea Shipping. *Seatrade Maritime News*, 30 May 2017. http://www.seatrade-maritime.com/news/europe/26033.html?highlight=InlhcmEi.

Hawaii New Network (HNN) 2019. This 132-foot Vessel Sailed from CA to Hawaii (and Back) without Anyone on Board, 15 February 2019. www.hawaiinewsnow.com/2019/02/02/16/this-foot-vessel-sailed-ca-hawaii-back-without-anyone-board/.

Hayashi, Y., N. Wakabayashi, T. Kitahashi, H. Wake. 1994. An Image Ranging System at Sea, Position Location and Navigation Symposium, IEEE, 1994, pp. 113–20.

ICS 2016. Manpower Report: The Global Supply and Demand for Seafarers. International Chamber of Shipping (ICS). Douglas W. Lang. ICS Conference, 7 September 2016.

Ifland, Peter. 1998. *Taking the Stars: Celestial Navigation from Argonauts to Astronauts*. The Mariners' Museum: Newport News, VA, 1998.

ITU-R M.824-4. Recommendation ITU-R M.824-4, February 2013. Technical Parameters of Radar Beacons.

Macaulay, George Cambell. 1890. *The History of Herodotus*, Vol II. McMillan & Co. London, New York, 1890. 2.5.28. www.vonsteuben.org/ourpages/hamanities/herodotus.pdf.

MAREX 2016. PMI Teams With Robert Allan on Autonomous Tug. *Maritime Executive*, 09 December 2016. https://maritime-executive.com/corporate/pmi-teams-with-robert-allan-on-autonomous-tug#gs.Ub2P87Q.

MAREX 2017a. BMT Awarded New Funding for Autonomous Navigation. *Maritime Executive*, 06 Decemebr 2017. https://maritime-executive.com/corporate/bmt-awarded-new-funding-for-autonomous-navigation#gs.FZmLCTE.

MAREX 2017b. Bourbon Joins Autonomous Ship Initiative. *Maritime Executive*, 07 November 2017. https://maritime-executive.com/article/bourbon-joins-autonomous-ship-initiative#gs.r9EWob0.

MAREX 2017c. Google and Rolls Royce Partner on Autonomous Ships. *Maritime Executive*, 03 October 2017. https://maritime-executive.com/article/google-and-rolls-royce-partner-on-autonomous-ships#gs.gPfAJ1k.

Meurn, Robert. 2013. *Anatomy of a Collision*. Tate Publishing, London, 2013. ISBN: 978-1-62746-820-6.

MSC 82/24. MSC 82/24/Add.2, Revised Performance Standards for Electronic Chart Display and Information Systems (ECDIS), ANNEX 24, p. 2.

MSC 98/20/2. Work Programme. Maritime Autonomous Surface Ships Proposal for a Regulatory Scoping Exercise. Regulatory Scoping Exercise for the Use of Maritime Autonomous Surface Ships (MASS). MSC 98/20/2, 27 February 2017.

MSC 99/5/1. Comments and Proposals on the Way Forward for the Regulatory Scoping Exercise. IFSMA and ITF, MSC 99/5/1, 22 February 2018, p. 3, para. 9.

MUNIN 2016. Final Brochure. The Project Maritime Unmanned Navigation through Intelligence in Networks (MUNIN), 2016. http://www.unmanned-ship.org/munin/.

Noris Dr., Andy. 2008. Integrated Bridge Systems Vol. 1, RADAR and AIS. The Nautical Institute, 2008, p. 45.

Patterson 2019. Ethiopian Airlines 737 MAX Pilots Followed Expected Procedures before Crash. CNN, 4 April 2019. www.cnn.com/2019/04/04/us/ethiopian-airlines-737-max-crash-preliminary-report/index.html.

Pillich, B. and C. Schack. 2002. Next Generation ECDIS for Commercial and Military Uses. *OCEANS '02 MTS/IEEE*, 2 (2002): 1025–32.

Promare 2019. The Mayflower Autonomous Ship. Promare. www.promare.org/,ayflower-autoship.

Ray, Tiernan. 2019. China's AI Scientists Teach a Neural Net to Train Itself. *ZDNet*, 22 January 2019. https://www.zdnet.com/article/chinas-ai-scientists-teach-a-neural-net-to-train-itself/.

Ronan, Colin A. and Joseph Needham. 1986. *The Shorter Science and Civilisation in China*. Cambridge University Press, Cambridge, 25 July 1986, pp. 28–9.

Sale 2017. Fulton's Steamboat Sensation. *New York Post*, 20 August 2017. https://nypost.com/2007/08/20/fultons-steamboat-sensation.

Sea Machines 2019. https://www.sea-machines.com/about.

Smith, A. and J. Anderson. 2014. *AI, Robotics and the Future of Jobs*. Pew Research Center, 6 August 2014. Sampled from views of Bill Woodcock (p. 51), Frank Pasquale (p. 53) and Gina Neff (p. 64). www.preresearch.org/wp-content/uploads/sites/9/2014/08/Future-og-AI-Robitcs-and-Jobs.pdf.

SOLAS 1974. Safety of Life at Sea Convention, 1974. Regulations V/19 and V/27, as amended.

Standards of Training, Certification and Watchkeeping (STCW) 2010. Manila amendments.

Statistica 2019. Number of Ships in the World Mmerchant Fleet as of 1 January 2018, by Type. *Statistica – The Statistics Portal*, 2019. www.statistica.com/statistics/264024/number-of-merchant-ships-worldwide-by-type/.

Sterling, Christopher H. 2011. *Encyclopedia of Military Communications From Ancient Times to the 21st Century.* Santa Barbara, CA: ABC-CLIO, 2011, p. 156.

Walker, Fred M. 2010. *Ships and Shipbuilders: Pioneers of Design and Construction.* Seaforth Publishing, Barnsley, UK, 2010. John Wilkinson (biography), p. 63. ISBN 978 1 84832 072 7.

Wingrove, Martyn. 2018. Alliance Unveils Autonomous Vessel Technology. Maritime Digitalisation & Communications, 16 July 2018. http://www.marinemec.com/news/view,alliance-unveils-autonomous-vessel-technology_53510.htm.

Wright, D. C. 1988. The Published Works of Sir George Biddell Airy. *Journal of the British Astronomical Association* 98, no. 7: 358.

Chapter 2

# Making the Case for Unmanned and Autonomous Ships

The motivation for autonomous commercial ships stems in part from a desire to reduce costs, enhance safety and decrease environmental risk associated with shipping operations. Improved competitiveness should be achievable for those taking advantage of the benefits of automation to create greater assurance of ship access by producing vessels that are less prone to failure with greater predictability through enhanced preventative maintenance strategies. Likewise, improved safety for mariners and safety of navigation for automated ships should be possible through the reduction or elimination of human error. However, these and many other such promises are often made by those without affiliation to the maritime industry and have yet to be proven.

There exists great momentum to look at the success of automation in other industries and to conclude the maritime industry should follow in their examples. However, questions remain as to whether the experiences of others are directly transferable to the maritime domain. For example, can comparisons realistically be made between autonomous automotive vehicle and ship operations? Does success in automated and remote operations of nuclear power plants assure similar results for Maritime Autonomous Surface Ships (MASS)? The answers to these and many other questions are being sought through experiments and trials in the use of automated ships and the application of automation technologies currently being undertaken using test beds and platforms at locations throughout the world.

A recent survey of over 600 members of the Institute of Marine Engineering, Science & Technology (IMarEst) provided valuable insight into attitudes of the shipping industry towards automation [Meadow, et al. 2018]. Nearly 25% of respondents claimed they would support a move to remote and autonomous shipping based on increased levels of safety, whereas 6% believed support would emerge from increased efficiency, and 9% emphasized reduced operating costs. However, 50% of the survey participants said support for remote and autonomous shipping would only come about from a mixture of all three elements [Meadow, et al. 2018, p. 8]. In response to a question regarding a move to the adoption of autonomous vessels 44% agreed that the shipping industry should follow the example of other industries such as the rail, mining and nuclear sectors that are successfully implementing and utilizing remote and autonomous operation. This was compared to 33% who disagreed that this was a solid basis from which to implement such a move; 23% remained neutral or did not answer.

A very interesting viewpoint was expressed in that ship owners and operators are generally skeptical as to the rationale for the significant investment necessary to move into automation and remote operation, especially in an economic climate that is challenging many sectors of shipping [Meadow, et al. 2018, p. 7]. The survey showed that more than 66% believed technology providers were the main drivers behind ambitions towards the adoption of commercial vessel remote and autonomous operation, and that less than 20% of respondents supported the view that providers of remote and autonomous technology fully understood shipping industry requirements. Conclusions reached included there appears to be a lack of a clear business case on which to base informed decisions regarding the establishment and implementation of MASS, especially as relates to cost reduction or increased revenue [Meadow, et al. 2018, p. 9]. The opinions articulated in the IMarEST survey are reflected in many other industry reports and articles that appear in recent literature.

A series of unscientific and unofficial surveys performed by the author at local watering holes where seafarers are known to congregate have had similar but more definitive results. Almost unanimously, the findings represent two distinct points of view. The first is a general opinion that no one particularly likes the idea of unmanned ships transiting areas that are congested and that any MASS large enough to be economically viable in trade would not have adequate maneuverability to avoid collision, especially in areas frequented by small watercraft. The second is an observation expressing consternation as to who really thinks MASS is a good idea anyway, especially since not one seafarer participating in the surveys had ever previously been asked directly, nor knows anyone who has been asked their opinion on the subject. This corresponds closely with IMarEST survey findings that outside, non-mariner technology providers are likely driving the movement towards automation. A small percentage of survey findings were eliminated from the results due to incoherence or other disqualifying condition on the part of the survey respondent.

This chapter seeks to examine these and other topics in greater detail to provide perspective as to the potential for success in development as well as benefits and risks that may result from unintended consequences of MASS implementation. The subject areas addressed include economic, safety and the environment. Other topics are considered within the confines of their own chapters in this book. The technologies involved in MASS related specifically to automation and autonomy are discussed in detail in Chapter 3, while MASS design and engineering is discussed in Chapter 4. The topics of command and control, navigation and communications are discussed in Chapters 5, 6 and 7, while security is discussed in Chapter 8 and training in Chapter 9. The various parties involved in the regulatory process at all levels are identified in Chapter 10, while legal instruments that may apply to MASS are discussed in Chapter 11. Finally, informed thoughts as well as speculation regarding the future directions of MASS are provided in Chapter 12. The information offered in these chapters will hopefully enlighten those interested in MASS of the factors most relevant to the subject and, through references cited, provide additional resources to further pursue their interest.

## 2.1 ECONOMIC PERSPECTIVES

Autonomous ships provide an entirely new platform upon which innovative business models can be developed to advance economical modes of transporting goods and materials between points, to leverage business opportunities through digitalization and to monetize data pertaining to shipping operations.

There appears to be a market for such innovation. The global autonomous ships industry is expected to generate $134.9 billion by 2030, with a compound annual growth rate of 4.4% from 2020 to 2030 according to a report published by Allied Market Research [Padalkar and Baui 2019]. This projection is based on an increase in demand for cargo transportation and a demand for increased operational safety of ships that will drive growth. According to the report the commercial segment will contribute the highest share of more than three-fourths of the total market share in the global autonomous ships market in 2020, and is expected to dominate during the forecast period. The passenger segment is expected to have the highest growth rate of 6.9% from 2020 to 2030. By region, Europe is expected to have the largest growth rate at 4.9% from 2020 to 2030 followed by North American at 4.4%. The Asia-Pacific region is expected to account for approximately half of the total market share in 2020, and is expected to register the highest revenue during this period.

### 2.1.1 Economical Transport

A common thread among nation states includes the encouragement of sustainable economic growth through increased trade and prosperity as exemplified by the United Kingdom, United States, China and others in their policy statements [Maritime 2050; USCG 2018; SCIO 2016]. The implication is that more ships will be needed to support the growth in trade. To stay competitive, all aspects associated with the non-recurring cost of ship design and production, the recurring costs for ship operations and the risks associated with the entire ship lifecycle must be re-examined on a continuous basis. Seafarer and human support alone can account for 30%–44%

of traditional ship costs in terms of salaries, crew quarters, bridge space, human interfaces and controls, and environmental systems (heating and air conditioning, food, water, lighting, plumbing, etc.), the elimination of which can potentially lead to significant cost savings [Minter 2017; CBI 2018]. Other considerations in terms of ship design and logistical support must also be taken into account.

The subject of ship automation is generally considered in terms of the benefits that may be achieved against the risks involved and the costs of accomplishing a meaningful return on investment. Part of this equation includes the viability of the technology used in the implementation of automation. The MUNIN project estimated a cost savings of $7 million over a 25-year period for an automated bulker versus a conventional bulker [MUNIN 2016]. However, detailed projections supported by verifiable facts are scarce. This is to be expected since the technologies needed for automation are still under development and the cost for production MASS is still substantially unknown. Advances presently being achieved in the field of machine learning and artificial intelligence for transport automation have brought the ship industry to the point where meaningful trials are currently underway to determine its feasibility and utility for MASS. With positive results coming out of these trials indicating great potential for success, this major stumbling block may soon be pushed aside to examine relevant issues in greater detail.

Ship propulsion using internal combustion engines is one area that is greatly dependent upon human supervision and interaction to perform routine lubrication and other tasks that seem to rule out fully automated vessels for use in long duration voyages over great distances such as trans-oceanic crossings. The use of highly reliable and more efficient electrical propulsion systems powered using batteries, alone or possibly in hybrid configurations with other methods of propulsion, would be most appropriate for short-sea shipping, ferries, harbor tugs and other functions over known transit routes with ready access to power and bunkering. However, internal combustion engines continue to become more reliable and efficient through advances in power management optimization and other techniques and opinions on this subject may change in future years [Jaurola 2018].

### 2.1.1.1 *Fixed Costs*

Fixed costs associated with ship design, development, test and acceptance along with the associated costs for investments and interest comprise extraordinary charges that occur early in the ship acquisition lifecycle as well as regular expenses for loan repayments and interest, registration fees, taxes, etc. Unmanned ships do not inherently require the features and systems necessary to support human life at sea; therefore, cost avoidance may be achieved by eliminating the non-recurring costs associated with their design, development, test and maintenance.

Such features include crew berthing compartments, galleys, heads, deck houses and other enclosed deck structures and their contents that are traditionally inhabited by and used for ship's crew and officers. Further cost avoidance can be achieved by the elimination of heating, ventilation and air conditioning, fresh water systems, sewage and holding tanks, interior electrical lighting, ship controls and human interfaces, entertainment and other support systems typically afforded to human activity.

The avoidance of costs for human features and support systems simplifies ship requirements and can make the design portion of the lifecycle shorter and less expensive than is normally the case. This will be offset somewhat by design considerations for autonomy-specific sensors and sensor systems as well as onboard computer and communication equipment needed to interconnect and transfer data and information internal to the vessel, reason with this information, and communicate decision-making results and supporting rationale to remote operations centers and other vessels. During development, further cost avoidance is achieved by eliminating those tasks associated with provisioning, fabricating and installing deck housing and the plumbing, electrical wiring and equipment, ventilation and other systems that normally would exist for conventional ships. Likewise, costs associated with the test and acceptance of these features and systems are also eliminated.

Additional benefits from the elimination of human support features and systems include larger cargo capacities, lower weight and the potential for more efficient hydrodynamics and less wind resistance than may be possible for conventional ships due to MASS no longer having a need for windows, doors and other portals exposed to the environment, with these economies resulting in lower fuel consumption and savings on fuel costs.

### 2.1.1.2 *Operating Costs*

Semi-variable costs associated with operating a vessel includes those incurred on a regular basis in the employment of the ship as well as the cost of assets necessary to support day-to-day vessel operations. These include crew wages, victualling, training, travel and repatriation; property, indemnity and hull insurance; regular maintenance including lubricating oil and spares, survey, cleaning, stores, administrative and management fees, and sundry costs. Provisions should also be made for capital expenses set aside for future expenditures such as periodic drydocking, extra survey and major equipment overhaul and replacement as well as the estimated costs for compliance with forthcoming regulations. Unless the vessel is chartered, additional operating costs associated with voyages undertaken would include bunkers, pilotage, tug hiring, port and canal charges and port agency fees as well as costs associated with loading and discharging cargo. Cost savings are also anticipated through lower fuel consumption, with electric propulsion becoming more widespread, assuming future electricity generation and delivery costs over national grids do not become cost prohibitive as consideration is given to phasing out the use of fossil fuels in preference to wind, solar and other green technologies.

The cost of operating autonomous vessels is anticipated to be less than that of conventional vessels as cost avoidance is anticipated in terms of onboard crew and pilotage, which will be offset somewhat by costs for voyage monitoring by remote operations centers. However, bridging the gap between a minimally manned vessel and a completely autonomous vessel is expected to present significant risk and may not be economically feasible in the short term. Additional savings are to be achieved through efficiencies gained from eliminating the maintenance costs of features and systems not required on automated ships as well as improved predictive maintenance and greater reliability in the various systems that are installed onboard these vessels.

## 2.1.2 New Business Opportunities

The advent of MASS provides support for innovation in the fundamental nature of shipping as well as managing the supply chain. In addition to the several MASS research and development test beds established in Norway, Finland, Singapore, China, the United States and other locations, maritime incubators focusing on various maritime technologies are being established in locations throughout North America as well as in Gdansk, Rotterdam and Mumbai [Hume 2018; PortGdansk 2016; PortXL 2016; BusinessLine 2016]. The Internet of Things (IoT) is providing many opportunities for the integration of sensors, systems and the vessels themselves to support vertical markets within the maritime industry and achieve gains in efficiencies and flexibility in MASS operations with resulting economic benefits from reduced costs. Much of this is expected to be achieved through real-time data and information gathering to aid in applications ranging from scheduling cargo shipments to vessel security, failure prognostics and ship maintenance, and other areas that will also result in new opportunities for businesses to collaborate in ways that were previously not possible.

However, the maritime industry is steeped in tradition and change is generally slow to occur. This often results in delayed adoption of technology, and MASS is no exception. According to one source their experience in interacting with the maritime market reveals that a relatively large number of organizations are not aware of the capabilities of technology today, the potential changes it could bring to the market, or the developments that have occurred in other markets because of technology [Kinthaert 2017]. Significant barriers to entering the maritime market exist based upon adherence to traditional business practices as well as overcoming regulatory obstacles to new technologies.

Examples of innovation are many. SailRouter B.V. uses artificial intelligence to measure ship motion in planning optimal ship routing using wind and currents to aid in navigation while maintaining minimal ship speed for on-time arrival at destinations [SailRouter 2019]. VesselBot has created artificial intelligence (AI)-based solutions to digitalize the chartering process providing operational, financial, and strategic benefits to users considering variables in real time to help match counterparties and negotiate transactions [VesselBot 2019]. Both of these examples illustrate the power of the IoT to acquire shipboard sensor data and combine it with weather, logistical and other information to create new solutions to old problems. The volume of data available through interconnected devices, systems and vessels can provide great insight at all levels of the maritime industry both afloat and shoreside. For example, Maersk technicians currently can pinpoint the location and operational details of any one of its 270,000 refrigerated reefer containers around the world [Murison 2018]. Airborne and undersea drones are increasingly being used for maritime inspection to detect problems on deck, in tanks and below the waterline, for security and surveillance, maritime search and rescue, and autonomous deliveries to ships [Karpowicz 2018]. These examples represent just a few of the opportunities seized upon by companies to create new products and services by harnessing IoT capabilities and AI technologies, with future potential bright for such ventures.

## 2.1.3 Data Monetization

The monetization of ship-acquired data has huge economic potential to gain operational efficiencies and also to expand revenue sources. To achieve automation, MASS will necessarily exist as one of the world's most sensor-intensive and instrumented platforms not only with respect to the vessel itself and its integrated systems, but also data on its cargos, the nature and quantity of other vessel traffic along its route of transit, the quality of aids to navigation with capabilities to automatically detect damaged and out-of-position aids that may endanger other vessel traffic, and countless other uses that are yet to be conceived. MASS networks can participate in multitudes of crowdsourcing initiatives, providing both meteorological data to enhance weather observations and oceanographic data concerning everything from fish and marine mammal observations to line and swath depth soundings that can supplement hydrographic survey efforts for navigation chart updates. MASS can also passively aid in law enforcement activities in much the same manner as public surveillance cameras. IoT data marketplaces wherein the trading and selling of not only the vast amounts of MASS collected data, but the information acquired through data and image analytics are largely unexplored territories with rich possibilities for new and innovative uses.

## 2.2 SAFETY

Human error is estimated to be responsible for between 76% and 94% of marine casualties [Allianz 2012; CBI 2018]. These statistics represent casualties affecting safety of navigation as well as the performance of shipboard duties by seafarers. The following paragraphs consider the potential effects of automated ships towards both of these topics.

### 2.2.1 Safety of Navigation

The rules governing safety of vessel navigation were formalized in the Convention on the International Regulations for Preventing Collisions at Sea (COLREG), 1972, whereby all vessels flying the flags of nation states ratifying the treaty are bound to the Rules [IMO 2003]. These regulations include general information pertaining to their application, the responsibilities of those involved and definitions of terms used within the regulations. They also contain specific regulations governing steering and sailing, lights and shapes, sound and light signals, exemptions, and several annexes providing additional details. Some countries also have internal regulations tailored to their specific needs and requirements that differ in one way or the other from the international rules and are embodied by Inland Rules applicable to that nation. The operation of MASS within unrestricted waterways in the presence of other vessels must be accomplished in compliance with the COLREG. However, IMO regulations as currently drafted preclude unmanned operations. The IMO is amending the regulatory framework, as discussed in Chapter 10, to enable the safe, secure and environmentally friendly operation of partly or entirely unmanned MASS and their interaction and co-existence with manned ships within the existing IMO instruments [MSC 98/20/2].

One of the greatest needs of MASS is to demonstrate the capability to operate correctly, consistently, reliably and successfully within the provisions of the COLREG. This is a complex task that, when unsuccessful, often results in incidents by both civilian and military vessels that fill the headlines of the world's media with reports on loss of lives, property and damage to the environment with great economic consequences. Clearly this is a challenging issue for even the most experienced of seafarers for which automated tools can make great contributions to enhance safety of navigation for all vessels. The challenge is even greater for MASS wherein automated reasoning capabilities must be trained to perform decision making with minimal or no human supervision. Moreover, preventing collisions is only one of the many skills needed for successful navigation. Details regarding the methods by which an onboard automated reasoning system may make the necessary decisions to perform vessel navigation are provided in Chapter 3 under the discussion of automation and autonomy. Further discussion is provided in Chapter 6 pertaining to the various sensors and sensor systems needed to provide the large volumes of data and information for an onboard reasoning system to accomplish navigation.

## 2.2.2 Job Safety

One of the many promises of automation is the potential to transform future jobs and the structure of the labor force. MASS is poised to make a significant contribution to reducing the potential for injuries and fatalities of seafarers, and especially deckhands, by detecting hazardous conditions, automating certain tasks and eliminating exposure to personal danger.

### 2.2.2.1 Hazard and Failure Prognostics and Detection

The advent of ubiquitous onboard smart sensors provides opportunities to view and assess, without human assistance, important deck and hold locations to determine whether any abnormalities or safety hazards exist in these areas, and to analyze the performance and condition of critical equipment and machinery for evidence of failure or pending failure. This includes tasks that are normally performed by deckhands as well as an entire new range of prognostic and diagnostic tasks amenable to electronic monitoring through enhanced built-in test and self-test capabilities integral to new system designs. Examples include visual sensing combined with multispectral image analytics that can determine whether lashings are fraying or becoming undone, when cargo may be shifting or is otherwise not secure, to detect and identify damaged ropes, wires and cables, and to visualize conditions over extended periods of time that may be indicative of the progression of failure processes such as smoke or dripping oil. Stress and vibration as well as many different types of smart sensors built directly into equipment can monitor and examine visual, physical, acoustical and electrical characteristics to search for symptoms of present and future maintenance issues, to detect the potential for failure and identify the early stages of failure so that corrective action can be initiated in a timely fashion.

A key element of failure avoidance is the availability of system redundancy combined with effective capabilities to detect and identify potential and actual failures and isolate failed equipment in a timely manner, and seamlessly reconfiguring equipment configurations to bring online independent, known-good equipment without interrupting vessel operation. This capability, where possible, must exist for all sensor, network, computer, communications, engine, steering, docking and mooring, power generation and distribution, firefighting, and onboard mechanical, hydraulic, electrical, electronic and other systems critical to vessel safety and performance. Furthermore, backup and redundant capabilities should be distributed geographically over wide areas of the MASS architecture with decentralized power and communications infrastructures to ensure continuous operation in the event portions of the vessel are damaged or destroyed due to collision, fire, flooding or other event.

Computer-aided design and engineering of vessel systems can now ensure critical functions and features may be identified and methods established for simulating their performance characteristics, followed by the monitoring of performance through the introduction of a wide range of sensors into the final implementation of the design. This may be accomplished through the implementation of functional and causal relationship analysis, failure mode effects and criticality analysis, and the identification and optimization of sensors capable of supporting these functions [Niculita et al., 2016].

### *2.2.2.2 Automation of Hazardous Tasks*

The deckhand's job is one of the world's most hazardous, with their having to work in the open in all kinds of weather, day and night on a moving platform and often with long working hours and little sleep. Work tasks include handling lines, wires and cables; securing cargo and deck machinery; standing watch in ships' bows, sterns and bridge wings to look for hazards to navigation and obstructions in the ship's path; assisting with docking and undocking, loading and unloading; and many other tasks to ensure overall vessel function and safety is maintained. The risks are great and include slips; falls, sometimes from great heights; being swept overboard; and the potential for being crushed by improperly secured machinery and cargo. Three examples are discussed:

*Enclosed Spaces* – One of the many hazards seafarers currently face includes entry into enclosed spaces onboard a vessel that may contain levels of oxygen that are incompatible with human life and/or flammable, noxious and hazardous gasses or vapors such as hydrogen sulfide and carbon monoxide. For many years safety alerts have been issued by the U.S. Coast Guard and other authorities documenting incidents leading to the deaths of mariners from these conditions including three experienced mariners while dewatering a mobile offshore drilling unit and suffocation of an engineer in an engine scavenging air receiver that demonstrate the risks involved [USCG 02-04, 04-19].

Enclosed spaces that are inadequately ventilated and are not designed for continuous occupancy may typically have small openings that provide entry and exit for inspection and other purposes. A solution to this nagging problem has gained momentum as a regulatory topic resulting in the issuance of a new regulation requiring the use of portable atmospheric testing equipment to test air quality prior to entry into the enclosed space [IMO XI-1/7]. Examples of such areas include cargo spaces; double bottoms; inter-barrier spaces; duct keels; pump rooms; chain lockers; voids; engine crankcases; engine scavenging air receivers; adjacent and connected spaces; and tanks used for fuel, ballast, sewage and liquid storage. The inspection of enclosed spaces is essential for periodic maintenance and to detect damage and potential leaks that may result from collision and other mishaps. This task becomes even more important on minimally or unmanned vessels where long periods of time may pass between inspections and the presence of backup personnel may not be available to assist should entry to enclosed spaces be required during a voyage.

The fitting of visual monitors to detect leaks, permanent sensors designed to detect traces of hazardous gasses and vapors, and even sampling the air in enclosed spaces using drones have all been considered in helping to solve this problem. The widespread implementation of such solutions throughout MASS enclosed spaces is essential to ensure vessel safety in the absence of properly trained seafarers to perform inspection tasks.

*Mooring and Docking* – Developments in the field of dynamic positioning have led to interest in automated docking system technology for vessels including MASS as a means to reduce human exposure to injury and potential damage from allision with both the vessel itself and the shore structures. During docking significant risk of injury or death exists from entanglement in lines and winches and in lines breaking and snapping back hitting someone. Injury to the back and muscles also occur as a result of repetitive tasks, manual lifting and handling of heavy lines needed as part of routine docking operations. The presence of fast-flowing waters and eddying currents due to tide and storm action along with variable winds and their effects on ship's hull and superstructure can challenge even the most experienced helmsman in performing precision docking maneuvers.

The processes involved in automatically docking a vessel along with the results of simulations using various operational scenarios highlight the complexities involved in this task [Bårslett 2018]. A demonstration of automatic docking technology was also accomplished by Wärtsilä in Norway with *Folgefonn*, a double-ended, 85-meter, 1,182-ton, hybrid diesel-electric powered passenger and car ferry [Farnsworth 2018]. Much of their development focused on the interfaces between environmental and propulsion systems to properly sequence thrusters as well as user-interface development to ensure proper operation. Automatic docking technology has now reached a level of maturity where its application to MASS can proceed forward with confidence. The main risk areas involve sensor fusion across large-scale vessels with multiple, geographically distributed viewpoint perspectives and the close integration of sensor and propulsion systems to ensure proper operation with minimal latency.

*Exposure to Hazardous Conditions* – Another example where risk to mariners can be reduced through task automation is in avoiding exposure to the dangerous conditions associated with firefighting and dealing with chemical hazards. The development of fixed and mobile robotic systems capable of entering hazardous areas and performing situation assessment, fire source detection, fire suppression, condition monitoring and other duties is well underway for shipboard, aerospace and land applications [Lattimer 2015]. Firefighting systems must be equipped with a large array of sensors to detect the presence of smoke and fire as well to assess the content and concentration of hazardous chemicals and explosive vapors in the environment. These sensors include flame, heat, smoke and chemical detectors corresponding to the materials used in the ship's construction and its contents including fuels and cargos.

## 2.3 ENVIRONMENT

Technology and data hold the key to a safer and more sustainable future, and, thanks to new technology emerging in fuel and energy use, automation and other areas, shipping is entering a new era [Lim 2018]. MASS represents a unique opportunity to implement the next generation of ship design using new technology that can project minimal ecological impact upon the sea on which they operate and into the air through which they pass. The full range of considerations needed to ensure MASS operations are environmentally sustainable must include their impact on the local environment in which they operate as well as their impact on the ecosystem as a whole.

Essential to accomplishing this goal is the need for MASS to ensure compliance with all environmental regulations and respect vicinities designated as Special Areas (SA) and Particularly Sensitive Sea Areas (PSSA) due to their unique or rare ecosystem, ecosystem diversity, or vulnerability to human activities to ensure damage does not occur as a result of their operations [IMO PSSA]. This applies to Marine Protected Areas (MPAs), Special Areas of Conservation (SACs), Sites of Specific Scientific Interest (SSSIs), Marine Conservation Zones and Marine Sanctuaries. MASS navigation must be accomplished in accordance with ship routing measures that include areas to be avoided, areas within defined limits in which navigation is either particularly hazardous or exceptionally important to avoid casualties and which should be avoided by all ships, or by certain classes of ships. Reliance on single methods of navigation such as Global Navigation Satellite System (GNSS) to operate relative to areas designated as SA and PSSA invites ecological disaster in an era of jamming and spoofing of satellite signals. Supplementary navigation systems such as inertial navigation, eLoran, virtual aids to navigation that require no physical infrastructure, and seabed contour following must be integrated into an overall navigation strategy to ensure the failure of any primary navigation system does not reduce the precision by which navigation must be conducted. Also essential is the need for situational awareness below the waterline using forward-looking navigation Sonar that will enable MASS to detect and avoid whales and other large mammals found ahead of the vessel along the path of transit.

The use of environmentally friendly technologies must also be addressed in MASS design in all areas to reach the goal of zero emissions to the external environment. This includes discharges into the sea related to ballast water, gray water, engine cooling and antifouling paint used to keep the hull clear of marine growth. Maritime shipping is also a significant contributor of greenhouse gasses into the environment, accounting for between

2.8% and 3.1% of annual emissions [IMO 2015]. However, concern over air emissions must not be limited to greenhouse gasses but also to air emissions of all kinds such as particulates and other compounds and gasses that are known to have detrimental effects on the environment.

Current designs for ship propulsion using internal combustion engines and inefficient drive trains are giving way to electric propulsion fueled with energy stored in batteries as well as other marine engines that rely on liquid natural gas, hydrogen fuel cells, wind power and hybrid propulsion methods comprised of two or more methods. Norway's fully electric and autonomous container ship *Yara Birkeland* is the first zero-emissions MASS at the point of operation as the power generation needed to charge its batteries is accomplished elsewhere. Reductions in fuel consumption are also anticipated due to more hydrostatically and wind-efficient MASS designs that may be streamlined without having concerns for operator visibility.

Research towards the development of antifouling materials and biocides presents significant opportunities to reduce or eliminate the toxic effects of these agents on the environment. However, the dilemma faced by industry is the fact that antifouling paints by design are intended to prevent the settlement, adhesion and growth of organisms to the painted surface, yet this must be accomplished in an environmentally friendly way. The application of copper in one form or another to ship bottoms has been the preferred method used for centuries. Large amounts of environmental data exist in the public domain regarding the long-term use of biocides that include Irgarol 1051 and diuron, and even dichlofluanid and zinc/copper pyrithione that provide a means to measure their effects on the environment. However, little public information exists for many of the newer biocides [Thomas and Brooks 2010]. More recent research has examined other agents and highlighted the need to develop suitable management practices for heavily contaminated biofouling waste to minimize their risk to soils [Ciriminna et al., 2015; Bighiu 2017].

**REFERENCES**

Allianz 2012. Safety and Shipping 1912–2012: From Titanic to Costa Concordia. Allianz Global Corporate & Specialty, March 2012, p. 3. www/agcs/allianz.com/PDFs/Reports.

Bårslett, Simon, Martin Longva and Thor-Inge Nygård. 2018. *Auto-Docking of Vessel*. Norwegian University of Science and Technology, Trondheim, Norway June 2018.

Bighiu, Maria Alexandra. 2017. *Use and Environmental Impact of Antifouling Paints in the Baltic Sea*. Stockholm: Stockholm University, 24 March 2017. ISBN 978-91-7649-692-3.

CBI 2018. Massive Cargo Ships Are Going Autonomous. CBInsights, 28 August 2018. https://app.cbinsights.com/research/autonomous-shipping-trends/.

Ciriminna, Rosaria, Frank V. Bright and Mario Pagliaro. 2015. Ecofriendly Antifouling Marine Coatings. *ACS Sustainable Chemistry & Engineering*, 3, no. 4: 559–65. DOI: 10.1021/sc500845n.

Farnsworth, Alexander. 2018. Look, Ma, No Hands! Auto-docking Ferry Successfully Tested in Norway, 27 April 2018. https://www.wartsila.com/twentyfour7/innovation/look-ma-no-hands-auto-docking-ferry-successfully-tested-in-norway.

The Hindu BusinessLine 2016. Huge Scope for Start-Ups in Maritime Sector. *The Hindu BusinessLine*, 20 January 2018. www.thehindubusinessline.com/economy/huge-scope-for-startups-in-maritime-sector-investors/article8442457.ece.

Hume, David. 2018. A Growing Blue Economy in North America. *Maritime Executive*, 27 May 2018. www.maritime-executive.com/blog/a-growing-blue-economy-in-north-america.

IMO 2003. Convention on the International Regulations for Preventing Collisions at Sea, *1972: Consolidated Edition 2003*. International Maritime Organization, London, 2003. www.imo.org/en/About/Conventions/ListOfConventions/Pages/COLREGS.aspx.

IMO 2015. *Third International Maritime Organization Greenhouse Gas Study. Section 3: Scenarios for Shipping Emissions 2012–2050*. International Maritime Organization, London, 2015, p. 18.

IMO PSSA. Particularly Sensitive Sea Areas, International Maritime Organization. www.imo.org/en/OurWork/Environment/PSSAa/Pages/Default.apx.

IMO XI-1/7. Guidelines to facilitate the Selection of Portable Atmospheric Testing Instruments for Enclosed Spaces as Required by SOLAS Regulation XI-1/7. MSC.1/Circ. 1477. International Maritime Organization (IMO), 9 June 2014.

Jaurola, Miikka, Anders Hedin, Seppo Tikkanen and Kalevi Huhtala. 2018. Optimising Design and Power Management in Energy-Efficient Marine Vessel Power Systems: A Literature Review, *Journal of Marine Engineering & Technology*, 2018. DOI: 10.1080/20464177.2018.1505584.

Karpowicz, Jeremiah. 2018. 4 Ways Drones Are Being Used in Maritime and Offshore Services. *Commercial AUV News*, 9 August 2018. https://www.expouav.com/news/latest/4-ways-drones-maritime-offshore-services/.

Kinthaert, Leah. 2017. 6 Maritime Startups That Are Changing the Game. KNect365, 17 April 2017. https://knect365.com/shipping/article/1149354e-68d9-4e74-9f91-a900ac869526/6-maritime-startups-that-are-changing-the-game.

Lattimer, Brian. 2015. Robotics in Firefighting. *Fire Protection Emerging Trends Newsletter*. Society of Fire Protection Engineers. Issue 100, 2015.

Lim, Kitack. 2018. IMO Secretary General. Speech given at the International Workshop on Maritime Autonomous Surface Ships and IMO Regulations. International Maritime Organization, London, 14 May 2018.

Maritime 2050. Department for Transport (UK). Maritime 2050: Call for Evidence. London, 2018, Para. 3.22, p. 25.

Meadow, G., D. Ridgwell and D. Kelly. 2018. *Autonomous Shipping: Putting the Human Back in the Headlines*. Institute of Marine Electronics, Science & Technology (IMarEST), Singapore, April 2018.

Minter, Adam. 2017. Autonomous Ships Will Be Great. Bloomberg, 16 May 2017. https://wwwbloomberg.com/opinion/articles/2017-05-16/autonomous-ships-will-be-great.

MSC 98/20/2. Work Programme. Maritime Autonomous Surface Ships Proposal for a Regulatory Scoping Exercise. Regulatory Scoping Exercise for the Use of Maritime Autonomous Surface Ships (MASS). MSC 98/20/2, 27 February 2017.

MUNIN 2016. Maritime Unmanned Navigation through Intelligence in Networks. MUNIN Final brochure. European Comm, 2016. www.unmanned-ship.org/munin/wp-content/uploads/2016/MUNIN-final-brochure.pdf.

Murison, Malek. 2018. The Maritime IoT Landscape in the Next Decade. Maritime Digitalisation & Communications, 6 August 2018. https://www.mariniemec.com/news/view,the-maritime-iot-landscape-in-the-next-decade_53814.htm.

Niculita, Octavian, Obinna Nwora and Zakwan Skaf 2016. Towards Design of Prognostics and Health Management Solutions for Maritime Assets. 5th International Conference on Through-Life Engineering Services, 2016. DOI: 10.10.16/j.procit.2016.10.128.

Padalkar, Pranav and Supradip Baui 2019. *Global Autonomous Ships Market: Opportunities and Forecasts 2020–2030*. Allied Market Research, August 2019. https://www.alliedmarketresearch.com/autonomous-ships-market.

PortGdansk 2016. The First Maritime Incubator in Poland will be Established in Gdansk. *Port Gdansk*, 10 November 2016. www.portgdansk.pl/events/the-first-maritime-incubator-in-poland-will-be-established-in-gdansk.

PortXL 2016. The Future of Maritime Innovation Starts Here. https://portxl.org.

SailRouter 2019. Rotterdam, The Netherlands. www.sailrouter.com.

SCIO 2016. *Development of China's Transport*. State Council Information Office, 29 December 2016. www.scio.gov.cn/zfbps/32832/Document/1537418/1537418.htm.

Thomas, K. V. and Brooks, S. 2010. The Environmental Fate and Effects of Antifouling Paint Biocides. *Biofouling*, 26, no. 1 (January 2010): 73–88. DOI: 10.1080/08927010903216564.

USCG 02–04. *Confined Space Entry. Marine Safety Alert 02–04*. United States Coast Guard. Washington, DC, 23 February 2004 and 14.

USCG 04–19. *Confined Spaces: Silent & Invisible Killers. Marine Safety Alert 04–19*. United States Coast Guard. Washington, DC, 12 April 2019.

USCG 2018. *Maritime Commerce Strategic Outlook*. United States Coast Guard. October 2018, 11.

VesselBot 2019. Athens, Greece. www.vesselbot.com.

# Chapter 3

# Autonomy, Automation and Reasoning

The concepts of autonomy, automation and reasoning are related in that they define goals, processes and a logical approach towards achieving the goals. These terms are discussed together in this chapter to distinguish between their differences and to provide context for their use as pertain to Maritime Autonomous Surface Ships (MASS). *Autonomy* refers to independence or freedom from human influence as a vessel proceeds from point A to point B. The different viewpoints in the maritime and other industries related to the various degrees of autonomy that may exist are discussed in the first part of this chapter. *Automation* refers to the use of automatic equipment in a system process or processes that make it possible for a vessel to proceed from point A to point B. Significant progress has been made over past decades in the development of automatic systems for engine control and monitoring, navigation, and other systems that comprise today's complex ships. These forms of process automation will establish the basis for MASS implementation and are discussed in Section 3.2. *Reasoning* provides the capability to coordinate shipboard automation processes to achieve the goals of autonomy. This involves MASS attainment of situational awareness at a resolution sufficient to capture all data and information necessary for proper decision making, command and control of system components essential to execute one or more courses of action to implement these decisions. The technologies needed to implement the decision-making processes to achieve autonomy are discussed in the third part of this chapter.

## 3.1 METRICS OF AUTONOMY

A good starting point for any discussion on autonomy would be to achieve consensus on the definition of the term prior to advancing regulatory development. Different interpretations exist as to the actual meaning of the term as well as its application and use. Background information is provided on the characterization of "autonomy" as expressed by various maritime interests at the beginning of the IMO regulatory scoping exercise.

The meaning of the term as used in vehicle autonomy, which is at present further advanced than MASS, is also given to provide perspective as to how the maritime industry may or may not align with the work of another industry attempting to achieve the same goal.

Consideration of the various concepts and definitions was made during the 100th session of the IMO Maritime Safety Committee (MSC) MSC100 in December 2018 as part of the regulatory scoping exercise and accord was reached to harmonize these different concepts for automation into a set of final definitions upon which regulatory measures would continue to proceed. A partial sampling of some of these different perspectives as well as the final IMO definitions adopted upon the conclusion of MSC100 is illustrated in Table 3.1. Details describing several positions on degrees of autonomy are given in the paragraphs that follow.

### 3.1.1 Lloyds Register

The Design Code for Unmanned Marine Systems produced in 2017 by the Lloyds Registry describes seven autonomy levels from 0 to 6 [Lloyds 2017]. *Autonomy Level 0* represents there are no autonomous functions performed during vessel operations and all actions and decision making are performed manually. This does not preclude onboard systems having some level of autonomy as long as there is a human in the loop controlling all actions. *Autonomy Level 1* indicates there is some level of onboard decision support where all actions are taken by a human operator. However, decision support tools using data provided by onboard sensors can present options or otherwise influence the actions chosen. *Autonomy Level 2* is where all actions are taken by a human operator but decision support tools can present options or otherwise influence the actions chosen. Data used by decision support tools may be provided by onboard systems and/or systems that are remote to the vessel. *Autonomy Level 3* indicates a human is actively participating in the loop supervising all decisions and actions that are performed. Again, data used in this process may be provided by onboard systems and/or systems that are remote to the vessel. *Autonomy Level 4* is where decisions and actions are performed autonomously with human supervision where high-impact decisions are implemented in a way to give human operators the opportunity to intercede and override these decisions.

*Autonomy Level 5* represents the fully autonomous condition where a human operator rarely supervises operations and decisions are entirely made and executed by the system. The greatest degree of autonomy is achieved at *Autonomy Level 6* where vessel operation is unsupervised and decisions are entirely made and executed by the system during the mission. A higher autonomy-level system may use a lower autonomy-level system as part of its control strategy and a complex system may be comprised of a combination of multiple systems operating at different levels.

*Table 3.1* Different Perspectives on MASS Levels of Autonomy

**Lloyds Register 3.1.1**

| | |
|---|---|
| Autonomy Level 0 | Manual |
| Autonomy Level 1 | On-board Decision Support |
| Autonomy Level 2 | On & Off-board Decision Support |
| Autonomy Level 3 | 'Active' Human in the loop |
| Autonomy Level 4 | Human in the loop, Operator/Supervisory |
| Autonomy Level 5 | Fully autonomous: Rarely supervised |
| Autonomy Level 6 | Fully autonomous: Unsupervised |

**Norwegian University of Science and Technology (NTNU) 3.1.2**

| | |
|---|---|
| Level of Autonomy 1 | Automatic Operation (remote control) |
| Level of Autonomy 2 | Management by Consent |
| Level of Autonomy 3 | Semi Autonomous |
| Level of Autonomy 4 | Highly Autonomous |

**Norweigan Forum for Autonomous Ships (NFAS) 3.1.3**

| | | | |
|---|---|---|---|
| Decision Support | Remote Control | | |
| Automatic | Automatic Ship | | |
| Constrained Autonomous | Constrained Autonomous | | |
| Fully Autonomous | Fully Autonomous | | |

**IMO MASS Autonomy Resulting from MSC100 3.1.6**

| | |
|---|---|
| Degree 1 | Ship with automated processes and decision support |
| Degree 2 | Remotely controlled ship with seafarers on board |
| Degree 3 | Remotely controlled ship without seafarers on board |
| Degree 4 | Fully autonomous ship |

**Maritime UK 3.1.4**

| | |
|---|---|
| Level of Control 0 | Human on board |
| Level of Control 1 | Operated |
| Level of Control 2 | Directed |
| Level of Control 3 | Delegated |
| Level of Control 4 | Monitored |
| Level of Control 5 | Autonomous |

**Society of Automotive Engineers (SAE) 3.1.5**

| | |
|---|---|
| Autonomy Level 0 | No Driving Automation |
| Autonomy Level 1 | Driver Assistance |
| Autonomy Level 2 | Partial Driving Automation |
| Autonomy Level 3 | Conditional Driver Automation |
| Autonomy Level 4 | High Driving Automation |
| Autonomy Level 5 | Full Driving Automation |

### 3.1.2 Norwegian University of Science and Technology (NTNU)

The Centre for Autonomous Marine Operations (AMOS) at NTNU developed a four-level representation of autonomy [Utne 2017]. *Level of Autonomy 1* provides for automatic operation by remote control where a human operator directs and controls all functions. Some functions at this level are preprogrammed and system states, environmental conditions and sensor data are presented to the operator through a human-machine interface. *Level of Autonomy 2* reflects management by consent where the system automatically makes recommendations for mission or process actions related to specific functions and prompts the human operator at important points in the process for information or decisions. At this level the system may have limited communication bandwidth and may exhibit time delay due to communication channel latency caused in part by physical remoteness. Many functions can be performed independently of human control when so delegated by the human. *Level of Autonomy 3* entails semi-autonomous operation or management by exception where the system automatically executes mission-related functions when and where response times are too short for human intervention. The human operator may override or change parameters and cancel/redirect actions within defined time lines and the operator's attention is only brought to exceptions for certain decisions. Highly autonomous operation occurs at *Level of Autonomy 4* where the system automatically executes mission or process-related functions in an unstructured environment with a capability to plan and re-plan the mission or mission processes. The human operator may be informed about progress but the system is independent and "intelligent", with the human essentially out of the loop. In manned systems the human operator is in the loop performing a more supervisory role and may intervene throughout the process.

### 3.1.3 Norwegian Forum for Autonomous Ships (NFAS)

The NFAS identifies four operational autonomy levels for merchant ships [NFAS 2017]. The *Decision Support* level corresponds to advanced ship types instrumented with anti-collision Radars (ARPA), electronic chart systems and common automation systems like autopilot or track pilots. The crew is still in direct command of ship operations and continuously supervises all operations. This level normally corresponds to "no autonomy". At the *Automatic* level the ship has more advanced automation systems that can complete certain demanding operations without human interaction, e.g., dynamic positioning or automatic berthing. The operation follows a preprogrammed sequence and will request human intervention if any unexpected events occur or when the operation completes. The Remote Control Centre (RCC) or the bridge crew is always available to intervene and initiate remote or direct control when needed. The *Constrained Autonomous* level is where the ship can operate in a fully automatic mode in most situations and has a predefined selection of options for solving commonly encountered problems such as collision and allision avoidance. It has defined limits to the options it can use to solve problems, e.g., maximum deviation from planned track or arrival time. It will call on human operators to intervene if the problems cannot be solved within these constraints. The RCC or bridge personnel continuously supervise the operations and will take immediate control when requested to do so by the system or when determined to be necessary. Otherwise, the system will be expected to operate safely by itself. The *Fully Autonomous* level is where the ship handles all situations by itself. This implies that a RCC or any personnel on the bridge will not be available at all. This may be a realistic alternative for operations over short distances and in very controlled environments. In a shorter-time perspective, this is an unlikely scenario as it implies very high complexity in ship systems and correspondingly high risks for malfunctions and loss of system.

### 3.1.4 Maritime UK

Six levels of control are identified in the Industry Code of Practice for Maritime Autonomous Surface Ships up to 24 meters developed by Maritime UK [Maritime UK 2017]. *Level of Control 0* encompasses a vessel that is controlled by operators aboard. *Level of Control 1* under operated control is where all cognitive functionality is performed by the human operator who has direct contact with the unmanned vessel, makes all decisions, and directs and controls all vehicle and mission functions. *Level of Control 2* reflects directed control where some degree of reasoning and ability to respond is implemented into the unmanned vessel. It may have capabilities to sense the environment, report its state and suggest one or several actions. It may also suggest possible actions, but the authority to make decisions is with the operator. The unmanned vessel will act only if commanded and/or permitted to do so. *Level of Control 3* is where the unmanned vessel is authorized to execute some functions. It may sense the environment, report its state, define actions and report its intention. The operator has the option to object to, veto or override intentions declared by the unmanned vessel during a certain time, after which the unmanned vessel will act. The initiative emanates from the unmanned vessel and decision making is shared between the operator and the unmanned vessel. *Level of Control 4* is the level where the unmanned vessel will sense the environment and report its state, defines actions, decides, acts and reports its actions. The operator may monitor the events. At the highest level of automation, *Level of Control 5*, the unmanned vessel will sense the environment and report its state, defines actions, decides, acts and reports its actions without (real-time) operator monitoring of events.

### 3.1.5 Society of Automotive Engineers (SAE)

In the course of furthering the adoption of automated vehicles the SAE developed a taxonomy describing a full range of six levels of driving automation in on-road motor vehicles [J3016 2018]. In paraphrasing this taxonomy, the lowest degree of automation is Level 0, *No Driving Automation*, which reflects the performance by the driver even when enhanced by active safety systems. Level 1, *Driver Assistance*, features execution by a driving automation system of either lateral or longitudinal vehicle motion control (but not both simultaneously) with the expectation that the driver performs the remainder of the driving task. Level 2 represents *Partial Driving Automation* where the driving automation system controls both lateral and longitudinal vehicle motion with the expectation that the driver performs object and event detection and response and supervises the driving automation system. *Conditional Driving Automation* is the third level where an automated driving system has the expectation that the driver is receptive to requests to intervene, including driving system and other vehicle systems failures, and will respond appropriately. Next is Level 4, *High Driving Automation*, with automated driving system and task contingencies without any expectation that a driver will respond to a request to intervene. The top level of automation is Level 5, *Full Driving Automation*, with unconditional performance by an automated driving system of the entire driving task and contingencies without any expectation that a driver will respond to a request to intervene.

### 3.1.6 IMO Definition of Autonomy

Upon the conclusion of MSC100 in December 2018 four degrees of autonomy were identified for the purpose of the scoping exercise [LEG 106/8/1]. *Degree One* features a ship with automated processes and decision support. Seafarers are on board to operate and control shipboard systems and functions. Some operations may be automated and at times be unsupervised but with seafarers on board ready to take control. *Degree Two* is a remotely controlled ship with seafarers on board. The ship is controlled and operated from another location. Seafarers are available to take control and to operate the shipboard systems and functions. *Degree Three* is a remotely controlled ship without seafarers on board. The ship is controlled and operated from another location. The highest level of automation, *Degree Four*, is a fully autonomous ship where the operating system of the ship is able to make decisions and determine actions by itself.

### 3.1.7 Comparisons between Different Approaches to Autonomy

All of the approaches considered for the establishment of measures of autonomy correlate fairly well at the lowest and highest levels of autonomy. The lowest represents completely manual operation by onboard mariners, possibly with some manner of automation from individual independent systems such as propulsion control. The highest level encompasses completely hands-off operation of all vessel functions by the automated vessel itself without human intervention, but possibly with human observation if so desired. The two through five intermediate levels of automation between these two extremes exhibit different levels of remote and onboard vessel operational control, with or without seafarers on board the vessel, with varying degrees of responsibility delegated between remote operations and onboard personnel.

The IMO model of four degrees of autonomy provides clear distinctions based upon the presence or absence of onboard seafarers and the establishment of remote control. It is significant that autonomy is not based upon specific definitions of technology, technical capacity or implementation as these may change over time. However, technical capabilities are implied as being less than comparable to human performance at the lower degrees and equal to or better than human performance at the greatest degree of automation. There is also an implication that remote control can be achieved in an ideal implementation that exhibits immediate processing capability with zero latency or propagation delay in the communication channel to provide real-time monitoring of MASS events. Compensation for such delay on the order of minutes and hours is routine during the journeys of interplanetary satellites that transit many thousands of miles of empty space between communications. However, the earthly need to respond to events in real time, especially in cases of collision and allision avoidance and in response to technical malfunctions, dictates that criteria be established to mitigate such effects and to transfer control to where decision making can be made most effectively. This could very well mean that MASS under IMO Degree Three Autonomy may have to rely on its own onboard reasoning capability to initiate a timely response to an event due to being unable to rely on and maybe even to override remote human intervention, to effectively deal with a situation. This begs the question of whether MASS should ever be operated at Degree Three if it is unable to or not qualified to revert to Degree Four when needed.

A significant and notable exception to these models is the SAE taxonomy for automated vehicles that has no provisions whatsoever for remote vehicle operation. This is wholly appropriate due to the shear improbability of simultaneous control and supervision, but not necessarily monitoring, of millions of independently operating vehicles. However, the demonstrated propensity for unplanned and nefarious hacking and even the takeover by remote control of automated vehicles through many documented incidents illustrates a gaping hole in the security protocols being used in the automotive industry [Reindl 2018]. The automotive industry is not alone with this problem as it has also been demonstrated in the aviation industry [O'Flaherty 2018]. This must serve as a stern warning to maritime operators not only as an indication of vulnerability for possible hijacking of

MASS, but the reality that attempts will be made to hijack these vessels and all necessary safeguards and preventive measures must be established and demonstrated to be feasible and effective during testing in controlled environments before being unleashed in an operational environment. This subject is discussed in detail in Chapter 8, "Security".

## 3.2 PROCESS AUTOMATION

MASS autonomy can only be accomplished through the automation of the many critical shipboard systems that provide the essential services and create the required data products needed to operate the vessel. These systems must be individually capable of performing their intended functions, and must also interact with other onboard systems and ships' crews as well as remote operators to ensure coordination of functions is achieved in carrying out a cruise plan.

Automation of shipboard functions related to propulsion control, power distribution and control, bridge watchkeeping, alarm monitoring, damage control and other purposes all have distinctly different requirements and represent unique challenges. As was noted in the previous discussion, the existence of one or more automated systems onboard a vessel does not necessarily imply greater degrees of autonomy. Even when aided with high autonomy seafarers must still interact with these systems as part of bridge watch keeping to assess their operational status, determine the meaning of key indicators, identify methods to apply previously defined strategies, and then implement these methods by coordinating multiple systems in response to the present situation. Discussion regarding select automated systems is provided in the paragraphs that follow. The topic of how automated reasoning systems can duplicate the role of the seafarer in MASS implementations with the potential for improved performance and enhanced vessel safety as well as safety of navigation is discussed in Section 3.3.

### 3.2.1 Propulsion Control

Watchkeeping for engine room and related systems has evolved to its present-day level featuring unmanned engine rooms with bridge control of all propulsion functions except when necessary to change filters and perform periodic equipment lubrication and in emergency situations when the crew must take over direct manual control. This includes oversight of the main drive including the engines and drive trains; propeller functions including azimuthing, pitch and speed control; thrusters, batteries and other energy sources; air compressors; separators, fuel oil and pumps; inert gas generators; and other necessary equipment and devices [Naukowe 2012]. The processes involve monitoring and regulating temperatures, flow rates, levels, viscosities, pressures and other parameters of system components within proper ranges for all intended operations that include start up, idling, cruising, maneuvering and shut down. In cases where hybrid propulsion systems are used, automation encompasses all aspects of monitoring and control for each independent propulsion system.

### 3.2.2 Power Generation, Distribution and Control

The production and distribution of electricity is essential to power shipboard machinery and compressors, equipment, instruments, sensors, navigation and all other ship functions. Electricity production systems involve a mechanical process using engine-driven electrical generators and alternators used to create electricity. This is then adapted for shipboard consumption using regulators, power conditioners, variable speed drives, static converters, and other electrical and electro-mechanical system components resulting in the ultimate production of regulated three-phase alternating current (AC) and direct current (DC) electricity. There can be a wide range of different voltages needed to power individual systems and devices onboard the same vessel. High-voltage systems are needed to operate electrical propulsion systems and thrusters can operate at 3.3kV, 6.6kV, 10kV AC and even higher voltages at frequencies of 50Hz, 60Hz or 400Hz [Balasubramanian 2017]. Low-voltage systems generally operate at 440V, 220V or 110V at frequencies of 50Hz, 60Hz or 400Hz.

The use of DC electricity onboard ships is generally limited to low voltage and current applications in the 12–24V range to power electronic navigation, sensors and communication devices, and is created locally to their points of need using power supplies and converters to change AC to DC power and to step down their voltages to the proper level. However, high-voltage DC power (i.e., >1,000Vdc) has been identified as a more efficient means of powering machinery and equipment by lowering fuel consumption, noise, vibration and emissions over similar AC-powered machinery [Alf Kåre et al. 2018].

Distribution of electricity for conventional AC systems must be performed efficiently throughout the vessel and is generally accomplished using separate generators and alternators for propulsion and service loads. In addition to the generators, each system contains main switchboards that act as the intermediary between the generator and the rest of the ship, bus bars that provide the means for transferring electricity, circuit breakers, transformers, wires and connectors to the equipment to be powered. Similar requirements exist for auxiliary and emergency power. Integral to these systems are automated controllers through which power distribution between power generation devices, the distribution apparatus and end-user equipment is regulated.

Power distribution automation entails the continuous monitoring and adjustment of the characteristics related to these systems to ensure optimal regulation and distribution and include mechanical (rotation speed, oil temperature and pressure, and water temperature), electrical (voltage and frequency), and power (rated and current load, maximum power and load, average power and power reserve). Similar characteristics are also monitored for DC systems; however, their complexity is less than that of AC systems since AC-related equipment is eliminated (e.g., transformers and variable frequency drives) and the monitoring of frequencies, electrical phases, harmonics and other electrical characteristics is not necessary [Kim et al. 2018].

### 3.2.3 Auxiliary Systems

Various degrees of automation exist with respect to vessel auxiliary systems and little commonality exists between them. Auxiliary system components are diverse in function, widespread in the nature of their physical and electrical interfaces and the work they perform, and can be geographically distributed across the entire vessel. They can include lighting and fuel systems, heating and air conditioning, heat exchangers, compressed air sources; fresh water, ballast, gray water and sewage control systems; and steering gear, winches, cargo handling and deck equipment. Many are capable of being controlled, managed and maintained from the bridge, and some may include up to full automation of some of their functions. However, initiation of their combined operations and oversight of their performance as a whole is generally a manual process.

This is one of the more challenging aspects of MASS design representing the "catch-all" category of everything that is not propulsion, power distribution or navigation within which many different requirements exist with equally complex systems to fulfill these requirements. Full automation of auxiliary systems is likely to initially require specialized development of custom equipment and interfaces to ship's systems. However, it also represents opportunities for the development of standards in these same areas that will benefit future shipbuilding efforts.

### 3.2.4 Navigation

Today's integrated bridge systems (IBS) often promote up to "one-man" bridge operation of many required bridge watch functions. However, there is no common definition for this term or for the scope of its responsibilities. Wärtsilä describes their IBS as a set of screens and modules allowing centralized access to navigational, propulsion, control and monitoring information [EMT 2019]. Sperry's IBS cites its ability to show images from the ship's Radars, electronic chart systems and conning information, as well as closed-circuit TV images [Sperry IBS]. The Kongsberg system features one-man control, rational route planning, automated route-keeping functions, fail-to-safe mode of operation, functionality and reliability [Kongsberg IBS]. These and other commercially available products all generally assist in the performance of tasks associated with ship guidance including heading and speed control, cruise routing, positioning, and other tasks through automated processes and by displaying navigation information to watchstanders for them to take appropriate manual action when necessary. More advanced systems also provide capabilities for dynamic positioning and automatic berthing.

The three IBS systems cited above each come with many options to tailor their capabilities to the needs of individual ships and missions. Some provide relatively few but necessary capabilities centered upon vessel navigation, position fixing and future course planning with the integration of relevant sensors including GNSS, depth Sonar, Radar, AIS, ARPA, ECDIS and related information. Others provide expanded capabilities that encompass tasks associated with propulsion control, alert notification and management, and the integration of imaging at critical locations around the vessel that cannot be easily seen from the bridge. These capabilities are made possible through the use of standard data buses and communication protocols whereby data is provided by sensors to one or more computer-based controllers, thereby establishing the basis upon which the automation of navigation tasks may be achieved. At present such automation is commonplace onboard many vessels, lightening the workload on watch standers and providing them with the necessary tools to perform their jobs more effectively. However, lookouts are very critical to the success of vessel navigation and the automation of this task is still in its infancy. Comprehensive automation of the navigation task has yet to be demonstrated that is capable of showing reasonable effectiveness in following the COLREGs, avoiding collisions and allisions, and operating within the constraints of waterways and port environments. However, the principles for autonomous surface vessels to navigate according to COLREGs are demonstrated [Woerner, et al. 2017].

### 3.2.5 Communications

Automation of communications between ships and land points is presently a routine accomplishment for many vessels that are in direct contact with RCCs. However, automation of communications services presently has a much broader scope that is useful to the vast majority of vessels. Digital Select Calling (DSC) is integrated into MF, HF and VHF radios to automatically transmit or receive alerts for distress, urgent or safety communications under the SOLAS Convention. For greater distances and even in remote locations, data and information exchange is accomplished through satellite services such as those provided by INMARSAT and Cospas-Sarsat. GMDSS also provides a means for sending and receiving distress signals via HF, VHF and satellite as well as sending and receiving maritime safety information via navigational telex (NAVTEX) messages. Emergency Position Indicating Radio Beacons (EPIRB) can be initiated both manually and automatically to provide position information for vessels as well as for individuals in distress at sea. The Search and Rescue Transponder (SART) sends and receives Radar signals in response to X-band Radars that is useful for determining locations of vessels in distress. Much of the communications is automated with the use of VHF radio, along with Radar, AIS and video monitoring to provide the data and information exchange needed for participation in Vessel Traffic Services (VTS).

Ever-expanding broadband connectivity via satellite links as well as a wide area network (WAN) will dramatically expand data and information exchange between ships and land operators as well as between the ships themselves. Of particular significance is the establishment of 5G communications via cellular and satellite networks that promises order-of-magnitude increases in bandwidth that will further promote opportunities for automation and remote control needed to achieve the successful implementation of MASS.

### 3.2.6 Alarm Monitoring and Damage Control

Many different types of alarms exist onboard vessels that monitor system health and identify performance anomalies and failures of ship's equipment, infrastructure and main structural components. The scope of this function spans all shipboard processes including propulsion, power generation, navigation, auxiliary systems and their constituent machinery and apparatus, as well as the hull and superstructure across all decks and levels.

Built-in test equipment and automatic test programs and procedures exist to detect out-of-tolerance and fault conditions for many onboard electrical and electronic devices and instruments. Electronic controllers of mechanical and electro-mechanical equipment also identify anomalies and failures that occur presently and in the near future. Reporting of failures is accomplished to a central alarm station where an operator can initiate decisive action to remedy the situation. Much of this process is also automated such that systems can be reconfigured using redundant equipment and spares to swap out defective equipment in exchange for known-good equipment, often without interrupting the critical processes themselves.

Alarms originating from fire and smoke detected within the vessel are in many cases already highly automated where their source(s) can be detected, localized, identified and isolated; cooling and smoke boundaries determined; smoke removal path(s) identified and removal initiated; and fire suppression deployed without human supervision. Likewise, hull breaches and flooding originating from collision and allision can also be detected and compartments isolated with the automatic closing of hatches and portals all the while sensing ship stability. Calculating the extent of the breach can also be automated along with determining appropriate and alternative courses of action as may be necessary in the present and future.

MASS automation requirements necessitate the identification of all alarm interface signals and their values along with real-time notification of controller actions, with supporting data and information used as the basis for decision making in alarm resolution.

## 3.2.7 Integration of Process Automation

The processes identified above all possess some degree of automation, many of which are quite sophisticated and fully capable of performing their functions with little or no human supervision. There is also a great deal of overlap spanning these systems with reasonable expectations that solution providers with different skill sets will have dissimilar perspectives on addressing issues of automation. At present the information they create is ultimately presented to a human operator in the form of recommendations and/or actions for concurrence and implementation either in consultation with or through the direct actions of human operators. For MASS the same scenario can be followed for remote operation through the timely communication of results and conclusions to a RCC where operators can perform the same role to handle events and mitigate problems. However, at higher levels such as IMO Degree Three and Degree Four Automation, onboard capabilities exhibiting multiple, higher levels of reasoning must emulate a human operator to digest these same data and information spanning all automated systems as a whole. Such capability presently has not yet been demonstrated as being possible or even feasible and is the focus of many present experiments and testbeds worldwide. This last step in achieving autonomy is envisioned to be accomplished through the use of artificial intelligence technologies that can perform to at least the same level as human operators under the same circumstances, and in many cases perform much better than humans. These technologies and their expected methods of implementation in MASS are discussed in the next paragraph.

## 3.3 MASS REASONING

The human ability to reason has provided the means to coordinate shipboard processes and to execute a voyage since the earliest days of sail. However, some people do this better than others and this has led to the development of teaching curricula, standards and regulations to guide the reasoning process to where humans can reasonably assume that certain actions will have predictable outcomes and to anticipate how others will respond in well-defined situations. The resulting establishment of maritime academies, systems of aids to navigation and rules of the nautical road has paralleled the development of technology to assist in learning the fundamental and practical lessons needed for proper ship operation; to provide tools to enhance decision making; and to enhance the ability to communicate between vessels and those on shore to gather information, convey intentions and minimize ambiguity.

Present-day technology has matured from the automation of simple, predictable tasks such as using a line to control a tow, to the use of a hydraulic winch incorporating load sensing to maintain constant pressure and make such processes safer and more efficient. Similar inroads have been made in automating many of the processes described in the previous section, with much of the work taken over by machines in lessening the dangers involved to enhance work safety and safety of navigation. This includes reducing the workload of seafarers where today's ships transport millions of tons of cargo safely and efficiently with a fraction of the crew previously required on board.

However, there is substantial room for improvement. Seafarers are still exposed to harsh working conditions exhibiting physical danger with comparatively little situational awareness in all kinds of weather across expanses of water capable of swallowing ships and their crews with little warning and leaving scant traces to determine what transpired. Human error and the results of poor planning and decision making fill the headlines with tales of sinkings, collisions, allisions, groundings, capsizes and many other outcomes that result in great loss of life and property, environmental disaster and ruined economies. Vessels bearing the names of *Exxon Valdez*, *Rena*, *Costa Concordia*, *Sewol*, *El Faro* and many others attest to such events that still happen all too frequently and that should have been prevented.

### 3.3.1 Some Thoughts on Reasoning

Dr. Charles Lamb, a neuroscientist at the University of California, San Francisco, has been quoted as saying that a fundamental attribute of what makes us human is the ability to respond to something we didn't anticipate happening, and that humans are hardwired to create [LaMotee 2019]. This is exemplified by seafarers who are renowned for their resourcefulness in dealing with many situations for which they have not been specifically trained, yet their fundamental education and experience allowed them to improvise and thus save their own lives and the lives of others and their ships.

Human reasoning is a complex subject about which many theories exist, and more so opinions. An oversimplified view seems to regard the logical approach as considering varying levels of abstraction that provides a conceptual and mathematical framework for analyzing information, while the psychological approach entails observation of the reasoning processes that embody empirical knowledge about what phenomena can be replicated and under what conditions [Stenning and van Lambalgen 2008]. In reality, lessons can be learned from both of these and other perspectives in defining approaches to automation. In looking at this from an experimental viewpoint, a particularly relevant issue to MASS is research into how the brain recognizes faces where portions of the cerebral cortex were identified as dedicated to this task, and the revelation of associated neural activity and their organization [Tsao 2019]. The results included derivation of quantitative measures for faces that comprise 50 features for shape (e.g., size, distance, lines and curves) and appearance (e.g., complexion, color and texture). This provides an example of how a face can be described and the responses in terms of how biological neuron activity can be measured when presented with individual faces and facial features. This same process and methodology may be adapted to represent other objects besides faces that may be experienced in the maritime domain such as ships and small boats, buoys, and land features such as a water tower or quay. Related approaches may also be used to identify patterns within data and to draw conclusions from the fusion of information acquired from multiple automated processes to emulate the higher levels of reasoning achieved by humans. The physical organization of neurons can also illustrate a mechanism that can be simulated using computer software and implemented in hardware to effect similar results as part of a reasoning capability that can reside on MASS and at a RCC in digital twin configurations to ensure consistency in operations as well as results in onboard processes.

There are also instances of bias and inaccuracies in human reasoning that must be considered and, if possible, compensated for in automated shipboard processes. Some of this can be caused by ignorance of things that are different from training examples and where specific examples were not considered or included. The architectures of reasoning systems must be capable of recognizing such bias exists so they can identify their presence in the reasoning processes and accurately represent them so they may be overcome to arrive at a satisfactory, if not optimal, solution.

Human education and training is accomplished in great measure by exposing people to the fundamental principles of the problem domain followed by examples of how these principles are applied to everyday situations likely to be encountered. The ultimate in human achievement is the ability to extend beyond one's education and experience to apply their knowledge and expertise to solve problems and overcome situations never before experienced. Methods for training shipboard reasoning systems must approximate these same principles to adapt existing and create new teaching and learning approaches to correspond with human capabilities. New methods must also be created that reflect the unique attributes of computer hardware and software to achieve these goals.

### 3.3.2 Artificial Intelligence

The search to improve human capabilities and to extend human influence further into the environment while minimizing risk continues to propel technology into emulating human thought and decision making. The overall goal is to accomplish tasks in a manner that is at least as good as is humanly possible, and to exceed human capabilities where able. We have invented devices that are physically stronger and faster; that provide us with better sight, hearing and smell; that have greater sense of touch; that are larger or smaller to fit into places; and that sense in ways that humans cannot. The next challenge is to cultivate the ability to use these potential advantages to allow us to reason about the environment and shipboard processes in such a way that enables us to do things and perform tasks that were previously impossible. The technological discipline that considers the solution to this challenge for MASS is known as artificial intelligence (AI).

The Association for the Advancement of Artificial Intelligence (AAAI) defines AI in terms of the scientific understanding of the mechanisms underlying thought and intelligent behavior and their embodiment in machines [AAAI 2019]. One of the most promising of AI technologies today is the field of artificial neural networks that simulate the neurons of the brain and drive human thinking processes. Their method of organization, simulation in computer software, implementation in computer hardware, and integration into onboard MASS reasoning systems that interact with automated processes and sensor systems are discussed in the paragraphs that follow.

### *3.3.2.1 Neural Network Architecture*

Artificial neural networks (ANN) consist of many individual neurons, or nodes, organized across multiple layers that are interconnected in various ways based upon the needs of the specific problem they are intended to solve. The layers can generally be categorized into three categories: an *input layer* consisting of multiple nodes representing the input variables, an *output layer* consisting of two or more nodes representing the solution or decision, and one-to-many internal or *hidden layers* containing tens to millions of nodes that are interconnected between themselves and the nodes of the input and output layers as illustrated in Figure 3.1. Interconnections between nodes represent the transform between the inputs and outputs as relate to associations of the problem and the solution as signified by if, when, where and how nodes are connected to each other, the strengths of their connections, and the existence of feedback and/or feedforward paths between layers.

There are hundreds of basic neural network architectures, many of which provide good solutions to problems in general, and others designed to solve specific classes of problems with precise solutions. These basic architectures must be tailored and adapted to the requirements of each individual application resulting in a custom network, the quality of which is entirely dependent upon the quality and strength of the training sets used in their creation.

Each MASS automated process has requirements that are unique in terms of its specific shipboard implementation. This refers not only to the distinctive configuration of each vessel on which they are installed,

Autonomy, Automation and Reasoning 53

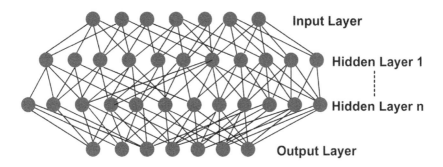

*Figure 3.1* Basic neural network architecture.

but also to the environment in which it operates. The interfaces between each automated process and the one or more system-level, cooperative MASS reasoning systems responsible for the vessel as a whole must be well defined and comprehensive to function effectively. The data and information that passes across these interfaces ideally must represent all nominal and potentially abnormal conditions likely to be encountered for each process. Likewise, the system-level MASS reasoning system to which the individual automated processes report must be knowledgeable of the complete range of findings and conclusions for each of the automated processes much like the Captain for each of the ship's departments. One or more auxiliary reasoning systems can assist the MASS reasoning system by providing specialized advice as the Executive Officer would advise the Captain. Interaction and coordination between automated processes is performed, with or without direct supervision by the reasoning system, as needed to ensure smooth operation. An example of these relationships and interfaces is illustrated in Figure 3.2.

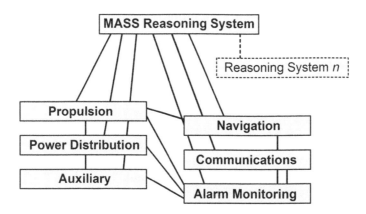

*Figure 3.2* MASS reasoning system interaction with automated processes.

Figure 3.1 and 3.2 each represents one of the many possible configurations of neural network architectures and reasoning systems that can be created to implement MASS. Several different approaches are being considered by the various projects identified in Chapter 1 and elsewhere that will produce results useful to their specific applications that may or may not be entirely practical to other initiatives. However, the lessons learned from each in the types of challenges and problems encountered and how solutions were achieved will be useful to many in the shipping industry and also to those involved in process and vehicle automation.

### 3.3.2.2 Software Simulation

The use of software to simulate the actions of biological and artificial neural networks has been for decades the predominant method for research and development in this field. This provides for relatively easy development, analysis and testing of many different experimental neural network architectures for accuracy of results and performance evaluation. Comprehensive specification and application of many process parameters and variations thereof can be evaluated as inputs to an onboard reasoning system in terms of data content and timing to gain understanding of process behavior and data variability and to optimize problem solutions. In addition to process automation, neural networks can also be used to mine exceptionally large datasets to search for patterns within the data that may be indicative of causal, effect and trend relationships encountered during vessel operations. Many software packages also provide tools to aid in visually monitoring progress in training as well as the architecture as it exists at any particular time and as it changes over time.

Inputs to neural networks can exist as data originating from processes, sensor signals existing in the time and/or frequency domains, and/or digital imagery of objects or events represented as pixels. The efficient processing of alphanumeric and sensor signal data can be aided using the fastest possible multiple core or parallel central processing units (CPUs) to perform the simulations. Rapid processing of imagery can be aided through the use of graphics processing units (GPUs) that are designed specifically to perform repeated operations using very large amounts of data using methods that would swamp traditional central processing unit (CPU) capabilities.

### 3.3.2.3 Network Training

The terms "supervised learning", "machine learning" and "deep learning" are all associated with AI, yet their significance is often blurred in terms of how these different approaches may or should be applied. Each of these terms represents a portion of what comprises artificial intelligence. The difference is simply this: supervised learning involves human guidance to teach a neural network the qualities used by humans in classification of decision making, while machine learning is a process whereby neural networks train themselves to learn a skill or task without human supervision and their performance improves with experience. Humans can guide and adjust machine learning processes in cases where an incorrect conclusion is reached or an inaccurate prediction is given. Deep learning is an extension of machine learning that enables a neural network to learn continuously and to detect and correct its own incorrect conclusions and predictions.

Supervised learning promotes relatively fast software development with reliable and consistent results in cases where a human can look at something and determine what exactly it is. For example, when training a watch standing system it is possible for a human to look at an image and specifically identify an aid to navigation as a water tower, buoy, light range or other such device. During supervised training a buoy can be identified by the human placing a circle or other annotation in the image around the object. In the case of a red, cylindrical (can) port hand mark in International Association of Marine Aids to Navigation and Lighthouse Authorities (IALA) region A, the object would be assigned to the class and subclasses *object* as "visual aid to navigation", *type* as "floating platform", *location* as "latitude" and "longitude" and characteristics as shape, size, elevation, color, lit/unlit, signal character, light intensity, sectors, construction materials, retro-reflective features, names, letters and numbers, as appropriate with suitable entries considering the IALA region (A or B) in which it is located [IALA 2001]. Supervised training of this nature can be successfully accomplished by one skilled in the subject of aids to navigation using several hundred to a thousand or more images or data patterns representing each unique instance of that type of object.

The use of machine learning to achieve the same goal involves the presentation of many images of buoys of the exact same type (but not instance) of red, cylindrical (can) port hand mark, only no human annotation or supervision is involved except for corrections of incorrect decisions or predictions. This type of training requires 5,000–10,000 or more images or data patterns of the type of object, with the neural networks themselves learning the correct object while continuously adjusting its internal node connections to improve its performance over time and with experience. Positive reinforcement (this IS the object), negative reinforcement (this IS NOT the object) and multiple object training (these ARE objects A, B and C; and, this IS object B but NOT A or C) may be used during training. Performance metrics will indicate continuous improvement over time for successful machine learning implementations.

Deep learning can be used to train a neural network using a process similar to machine learning, but entirely without human supervision or intervention in the case of an incorrect answer or prediction. A deep learning neural network has a decidedly more complex architecture with more levels and nodes per level than used for machine learning in the same task. Deep learning requires tens to hundreds of thousands of images or data patterns representing the object. The requirement for such large numbers of training data has, until recently, posed a significant obstacle in the successful implementation of true deep learning projects. However, the successful creation of a new generator architecture and its use in a machine learning generative adversarial network by NVIDIA to produce thousands of highly detailed AI-generated faces that do not correspond to any specific person provides great hope in the near future to automatically create highly accurate training sets in sufficient numbers to accomplish deep learning in this large scale [Karras, et al. 2019].

Researchers at the Georgia Institute of Technology (Georgia Tech, Atlanta, USA) have developed a prototype programmable integrated circuit that efficiently solves a large class of optimization problems useful to speed up and enhance the training of neural networks along with other applications [Chang, et al. 2019]. This 49 optimization processing units (OPU) core integrated circuit using an alternating direction method of multipliers (ADMM) architecture is 4.77 times more power efficient and 4.18 times faster than a comparable graphics processing unit (GPU). It breaks complex problems into smaller problems, works on each and comes up with the best overall solution, and shares the results. Such technology can ultimately benefit many machine learning applications that presently require significant amounts of time to develop.

### 3.3.2.4 Hardware Implementation

Trained neural networks can be deployed into operational reasoning systems in their software form. However, for faster performance as may be required for decision making in real time the software can be implemented with hardware assistance using a neural processing unit (NPU) that is optimized to perform neural computing. NPUs are analogous to and supplement a multicore CPU or parallel CPUs to achieve near real-time computing performance. Furthermore, the use of GPUs as described in the previous paragraph to supplement CPU and NPU assets for real-time processing will result in close to optimal performance that will improve over time as these processing units mature.

High-performance computing featuring highly integrated CPU, GPU, OPU and NPU architectures can be built for research and development purposes but are not generally available in sizes and configurations that would be useful for MASS implementation. However, their present use in MASS test beds and the integration of such future systems into automated reasoning systems for operational MASS is expected to be a forerunner of the type that will fulfill computational requirements through the 2020s.

## REFERENCES

Alf Kåre, Ådnanes, Ismir Fazlagic, Jan-Fredrik Hansen and Alf Kåre Ådnanes. 2018. Onboard DC Grid – One Year in Operation. ABB Asea Brown Boven, Ltd., 2018. https://new.abb.com/marine/generations/technology/onboard-dc-grid-one-year-in-operation.

Association for the Advancement of Artificial Intelligence (AAAI) 2019. Definition Found on AAAI Home Page. www.aaai.org.

Balasubramanian, Sriram. 2017. Reasons for Using High Voltage Systems On board Ships. Marine Insight, Marine Electrical, 31 May 2017. https://www.marineinsight.com/marine-electrical/reasons-for-using-high-voltage-systems-on-board-ships/.

Chang, M., L. Lin, J. Romberg and A. Raychowdhury. 2019. OPTIMO: A 65nm 270MHz 143.2mW Programmable Spatial-Array-Processor with a Hierarchical Multi-cast On-Chip Network for Solving Distributed Optimizations, Georgia Institute of Technology. *IEEE Custom Integrated Circuits Conference*. Austin, Texas, April 2019.

EMT 2019. Integrated Bridge Systems. Wärtsilä Encyclopedia of Maritime Technology, 2019. https://www.wartsila.com/encyclopedia/term/integrated-bridge-systems-(ibs).

IALA 2001. Aids to Navigation Guide. *NAVGUIDE*, Edition 4. International Association of Marine Aids to Navigation and Lighthouse Authorities (IALA), Monaco, December 2001, p. 31.

J3016 2018. *Taxonomy and Definitions for Terms Related to Driving Automation Systems for On-Road Motor Vehicles*. Surface Vehicle Recommended Practice, SAE International, J3016, June 2018.

Karras, Tero, Samuli Laine and Timo Aila. 2019. A Style-Based Generator Architecture for Generative Adversarial Networks. NVIDIA, 29 March 2019. https://arxiv.org/pdf/1812.04948.pdf.

Kim, Kyunghwa, Kido Park, Gilltae Roh and Kangwoo Chun. 2018. DC-Grid System for Ships: A Study of Benefits and Technical Considerations. *Journal of International Maritime Safety, Environmental Affairs, and Shipping*, 2, no. 1: 1–12. DOI: 10.1080/25725084.2018.1490239.

Kongsberg IBS. Integrated Bridge System. Kongsberg. https://www.kongsberg.com/maritime/products/bridge-systems-and-control-centres/navigation-system/integrated-bridge-system/.

LaMotee, Sandee. 2019. Jazz up your Brain, a quote of Dr. Charles Limb. Reader's Digest, February 2019, p. 119.

LEG 106/8/1. Outcomes of MSC 99 and MSC 100 regarding MASS Note by the Secretariat. Regulatory Scoping Exercise and Gap Analysis of Conventions Emanating from the Legal Committee with respect to Maritime Autonomous Surface Ships (MASS). Legal Committee. 106th session. Agenda item 8, 11 January 2019.

Lloyds 2017. Design Code for Unmanned Marine Systems. *ShipRight, Design and Construction*. Additional Design Procedures, Lloyds Register, February 2017, pp. 1–2.

Maritime UK 2017. Maritime Autonomous Surface Ships up to 24 Metres in Length. Being a Responsible Industry. An Industry Code of Practice: A Voluntary Code. Version 1.0, November 2017, Table 2, p. 13.

Naukowe, Zeszyty. 2012. Automation of Ship and Control. *Scientific Journals of the Maritime University of Szczecin*, 30, no. 102: 133.

NFAS 2017. Definition for Autonomous Merchant Ships. Norwegian Forum for Autonomous Ships (NFAS). SINTEF Ocean AS. Rev. 1.0, 10 October 2017, Table 1, p. 12.

O'Flaherty, Kate. 2018. How To Hack An Aircraft. *Forbes Magazine*, 22 August 2018. https://www.forbes.com/sites/kateoflahertyuk/2018/08/22/how-to-hack-an-aircraft/#6e2180c741d1.

Reindl, J. C. 2018. Car Hacking Remains a Very Real Threat as Autos become Ever More Loaded with Tech. *Detroit Free Press*, 14 January 2018, updated 15 January 2018. https://www.usatoday.com/story/money/2018/01/14/car-hacking-remains-very-real-threat-autos-become-ever-more-loaded-tech/1032951001/.

Sperry IBS. Integrated Bridge System Technology. Sperry Marine. http://www.sperrymarine.com/integrated-bridge-system/technology.

Stenning, Keith and van Lambalgen, Michiel. 2008. *Human Reasoning and Cognitive Science*. MIT Press, Cambridge, MA, pp. 3–4.

Tsao, Doris Y. 2019. Face Values. *Scientific American*, February 2019, pp. 22–29. www.scientificamerican.com.

Utne, Ingrid Bouwer. 2017. *Shipping and Digitalization 2013–2022*. Norwegian University of Science and Technology (NTNU) Center for Autonomous Marine Operations and Systems, 18 October 2017, p. 18.

Woerner, Kyle L., Michael Novitzky, Michael R. Benjamin and John J. Leonard. 2017. *Legibility and Predictability of Protocol-Constrained Motion: Human-Robot Ship Interactions under COLREGS Collision Avoidance Requirements*. Massachusetts Institute of Technology, Cambridge, MA. MIT-CSAIL-TR-2017-009.

Chapter 4

# MASS Design and Engineering

Topics associated with the design and engineering of Maritime Autonomous Surface Ships (MASS) span all shipboard disciplines and beyond the vessel to also encompass shoreside and remote support functions. This chapter begins with a discussion of the missions, operational settings and environmental factors within which these vessels are anticipated to operate as well as hull and deck design as may be peculiar to MASS, and propulsion to fulfill the precise maneuvering requirements associated with remote and fully autonomous operations. The various onboard sensors are also discussed including sensor types, their effective placement and issues to be considered to help maximize their usefulness and minimize interference caused by other systems. Some thoughts are also given regarding factors essential for maintenance pertinent to unmanned vessels.

The second part of this chapter describes 14 different initiatives for the development of remote-controlled and autonomous vessels planned and in actual operation to illustrate current advances in MASS technology.

Finally, the last part of this chapter considers how port operators may create shoreside facilities, harbors and approaches to provide services necessary to accommodate the specialized needs of MASS.

## 4.1 APPLICATIONS AND OPERATIONAL SETTINGS

A wide range of circumstances and conditions are potentially suitable to MASS operations for which appropriate design and engineering disciplines must be determined to implement this concept. Several different applications have been identified as potential candidates for MASS development including ferries, short sea shipping, offshore support, tug, reconnaissance, firefighting and other duties. Other considerations include the locations where they will be used such as inland, near coastal or open ocean environments.

### 4.1.1 Hull and Deck Design

An ideal hull configuration would promote steering a straight course in calm and rough seas, exhibit great stability at all speeds and provide high maneuverability to enable rapid changes in course to take evasive and all other desired actions. As with all ships, hull designs for autonomous vessels represent a series of tradeoffs between speed, stability and maneuverability as warranted to ensure safe operation under the environmental conditions within which they are expected to operate. However, significant room exists for new and smaller autonomous vessels to supplement traditional vessel types in expanded roles that include resupply and observation.

Few anticipate initial hull designs will differ significantly between MASS and crewed vessels that serve the same functions and operate under similar conditions and geographical locations. For example, ferries, container ships, bulkers, offshore support and other such vessels exhibit decades of purpose-driven design, build knowledge and experience to optimize the carrying of passengers and roll-on roll-off (RO-RO) vehicle traffic, containers, cargos and equipment. Likewise their hull forms consider whether they operate within inland and near coastal settings or if they will be exposed to the extreme conditions likely to be encountered in open ocean crossings. Progress continues in the development of new concepts that will help in making hulls more efficient such as the use of bubbler systems to reduce friction and improve performance, and there is an assumption that such advances may ultimately be beneficial to MASS and crewed vessels alike. However, the higher-level of instrumentation and hull sensors likely to be integrated into MASS designs as well as through-hull access and other facilities required for remote and autonomous undersea vehicles (RUV/AUV) deployment from MASS may ultimately result with hull designs diverging in the future. Bubbler systems, for example, may interfere with hull sensors needed to achieve comprehensive situational and environmental awareness. One area that might require modification of the hull is the installation of equipment and fixtures to support automated docking as may be needed by magnetic and other such systems devised in the future.

Deck designs, however, are another matter and there are specific characteristics of MASS that will ensure differences exist between fully autonomous and crewed vessels. For example, bridge structures can be minimized since there will be no need to accommodate human habitation or for the extensive use of ports designed specifically for human visibility and hatches for routine human access while underway. Likewise, crew sleeping compartments, galleys, heads and other spaces normally used by crews as well as the supporting plumbing and tankage, electrical, heating, air conditioning, ventilation and other support systems can be minimized or will not be needed at all. The deck and holds must be suitable for automated loading and unloading of containers and cargo as well as for line handling as may be needed for berthing and unberthing. A small hangar will also be needed at a point high on the deck to house and support one or more small remote and autonomous aerial vehicles (RAV/AAV) that will extend specialized inspection and observation capabilities and environmental situational awareness.

### 4.1.2 Propulsion and Power Generation

One of the biggest limitations in the implementation of MASS for transocean and other long-distance routes has to do with the physical characteristics of internal combustion engines and their need for constant lubrication and maintenance of parts and drive trains to ensure continued reliability. New requirements are also being established for the use of low-sulfur fuels in general and to reduce emissions in ports. These factors are the main driving forces behind new and green technologies for ship propulsion that include alternative engine designs and propeller systems that provide much greater reliability and maneuverability at lower cost, and new fuels and power sources that promote low or no vessel emissions.

#### *4.1.2.1 Engines*

The traditional means of propelling ships through the water since steam power that was beginning to be phased out in the middle of the 20th century has been the two-stroke, low-speed diesel internal combustion engine burning heavy fuel oil. These were typically connected directly to a propeller through a shaft. Running at constant speeds up to 100–150 revolutions per minute (RPM) these engines, when compared to other means of propulsion, were fairly easy to maintain and relatively inexpensive to operate but at a high cost in heavy air pollutants emanating from their exhaust stacks. Four-stroke diesel engines operating at higher RPMs connected to the propeller through reduction gears operate over a wider range of speeds and, in conjunction with gas turbines, can operate more efficiently and with greater power to propel ships at higher speeds.

Today's use of diesel-electric generators to power electric motors connected to both fixed and azimuthing propeller pods capable of rotating through 360 degrees as well as bow and stern thrusters provide better steering and maneuverability and form the basis for propelling some MASS implementations. In view of the IMO 2020 global sulfur limit for ships operating outside designated Emission Control Areas, challenges exist to come up with engines amenable to the use of alternatives to heavy fuel oils as well as lighter, low-sulfur fuels [IMO PAPS]. These include combinational solutions involving hybrid systems of diesel-electric and battery power as well as other engines burning different types of fuels such as liquid natural gas (LNG).

The use of dual fuel diesel-LNG engines produced by Wärtsilä, Yanmar, MAN, Caterpillar and other manufacturers is seen as a means to meet the new IMO 2020 emission limits through retrofit of existing systems as well as for installation in new-build vessels. Hydrogen burning engines are also currently being developed by Hyundai Heavy Industries that are envisioned to have even fewer emissions [Oh 2018]. Electrically propelled vessels using battery power create no emissions at the point of use, with the power used to charge the batteries produced elsewhere and transmitted to the vessel via an electrical grid.

### 4.1.2.2 Fuel and Power Sources

According to Lloyds Register Marine heavy fuel oil, the primary maritime fuel since the 1950s, will continue to be the dominant fuel for the majority of large vessels that include container ships, bulk carriers and general cargo, and tankers through 2030 with approximately 47%–66% of the market, down from around 80% in 2010 combined with marine gas oil, marine diesel oil and low-sulfur fuels to comprise the largest segment of the overall market [GMFT 2030]. LNG will gradually be adopted to increase usage from less than 1% in 2010 to in excess of 11% by 2030. The remaining fuels including methanol, hydrogen and battery power will account for the remainder of the market.

An example of a vessel using hybrid LNG and diesel propulsion is *Viking Energy*, an offshore supply ship owned by Eidesvik [Equinor 2018]. The vessel was built in 2003 measuring 5,073 gross tons, 6,013 tons deadweight (95m length ×20.5m breadth with 6.4m draft, 311ft × 67ft × 21ft) with a maximum speed of 17 knots and an economical speed of 10 knots [MarineTraffic 2019a; Eidesvik 2019]. In 2015, Eidesvik installed a 653 kWh/1,600 kW battery for propulsion during transit to absorb peak power loads by quickly contributing additional power when needed during demanding operations. It replaced one of the main engines as a "spinning reserve" when operating in dynamic positioning (DP) mode using GNSS and computers for precise positioning. The initial goal was to cut emissions, and fuel consumption was also reduced by 16%–17%. However, rather unexpectedly, they realized a 28% reduction in fuel consumption and emissions when operating in DP mode. Safety has also increased through better maneuverability where the generator does most of the work in supplying electrical power and the battery absorbs the variations up and down, resulting in even sailing without needing great surges of engine power to cope with waves.

Battery-powered vessels include those with relatively short, regularly scheduled routes such as ferries and shortsea shipping. MASS such as *Yara Birkeland* and the Seashuttle project operating in these roles are described in the second part of this chapter. The Ulstein plug-in hybrid ferry *Color Hybrid* also uses full battery power for ports of its journeys [Ulstein 2019]. One area where the use of batteries can be improved is in the adoption of technologies and practices that will permit battery regeneration using kinetic energy generated by slowing the vessel, wave action and other means. Batteries manufactured from advanced materials such as silicon carbide and gallium nitride also show potential for increased efficiency and effectiveness in battery performance. Another benefit from batteries is the potential to place their weight low in the hull to reduce or eliminate the need for water ballast.

Hydrogen fuel cell use on MASS is an option being considered. The results of its use on conventional vessels on an experimental basis can also transfer directly to remotely controlled and autonomous vessels. One example is the ferry *Water-Go-Round*, the first fuel-cell-powered vessel produced in the United States, expected to service the waters of San Francisco Bay in autumn 2019 [FCHEA 2018]. The 70-foot-long aluminum catamaran will have a top speed of 22 knots with 84 passengers and will be able to store enough hydrogen to run for up to two days between refueling. Built by Bay Ship and Yacht Co. of Alameda, California, it features dual electric motors from BAE Systems and a set of fuel cells from Hydrogenics. Refueling will be accomplished from hydrogen-carrying trucks without requiring any special dock-side infrastructure.

Another potential fuel is green ammonia produced using renewable energy with the potential that no greenhouse gasses would be emitted at any point in the production or use lifecycle [Ash & Scarbrough 2019]. A potential alternative to hydrogen, green ammonia already has existing global infrastructure as part of the fertilizer production process, does not require cryogenic storage and is a relatively energy dense liquid that would provide for efficient energy storage. An additional benefit is that standards and procedures already exist for managing ammonia.

Wind turbines are another potential source of power for propulsion and to recharge batteries. The Swiss company Bonum Engineering and Consultancy (Baar) and the Finnish shipping company Bore (Helsinki) are collaborating on the development of vertical wind turbines as a renewable energy source for ships [MAREX 2019]. Bonum Engineering and Consultancy will supply vertical wind mills for a Bore vessel to validate the technology. Maersk Tankers has also been testing this concept using 30-meter tall metal rotor sails on the product tanker vessel *Maersk Pelican*, where reductions in fuel consumption of up to ten percent are expected [Gronholt-Pedersen 2018].

The use of wave foils at the front of a ship to harvest the energy of waves to help propel a ship forward is another means to save energy and reduce fuel consumption as well as to dampen pitching and heaving motion while underway [Brandslet 2019]. Wavefoil (Hofstad, Norway) has developed a six-ton heavy-duty retractable module that is fitted into the forepeak tank of the Strandfaraskip Landsins transport company's 45-meter-long ferry M/F *Teistin*, whose route features rough Atlantic weather. Fuel savings are predicted at about four percent along the coastal route between sheltered islands from Bergen to Kirkenes, with savings of up to 15% possible where there are more waves.

### 4.1.3 Sensors

The key criteria for the successful accomplishment of MASS are the complement of sensors and the data and information they provide to:

1. Enable remote operators and onboard reasoning systems to visualize the physical environment in a manner that meets or exceeds that viewed by a human onboard the vessel performing traditional seafarer roles, and
2. Comprehend overall situational awareness that facilitates good, proper and correct decision making to ensure success in carrying out a safe voyage.

The present state of technology is such that it is now impossible to meet either of these criteria. However, we are getting closer as existing technologies become more reliable and new technologies provide greater capabilities. There still remains the need for better methods to be developed to achieve greater traceability between requirements and smart sensor design implementations, as well as for improved sensor system verification, validation and testing.

#### 4.1.3.1 Environment Visualization

The first criterion considers the capacity to overcome human limitations by broadening the extension of human senses beyond traditional physical and conceptual barriers, the technical performance of sensors to acquire data, and the ability to create useful and actionable information based upon these data. Ubiquitous sensing is rapidly approaching the state where physical barriers are no longer a problem. Sensors can be placed almost anywhere with no or low latency to promote real-time reporting of events, and even more so on ships where a multitude of data and imagery sensors can be placed at critical locations that are well known and have long plagued seafarers without adequate watchstanders. The subject of sensor placement on ships is discussed later in this section. However, conceptual barriers have yet to be overcome so long as critical sensor data is reported in forms and formats that are inconsistent, incomplete and lacking interoperability between systems to ensure these data may be interchanged without loss of fidelity and meaning. This problem has been solved for many traditional maritime sensors where data is reported on standard electrical and communications channels (e.g., NMEA 0183 and 2000). However, new sensor types reporting non-traditional subjects and topics such as image sensors designed for specific purposes that include boat and buoy recognition are still in their infancy with custom systems being created and little coordination between developers. Such problems are not insurmountable, but forward progress will be slowed and the same mistakes repeated by different organizations until consensus may be reached on these concepts.

The technical performance of today's sensors is making giant leaps in areas such as range of coverage (e.g., visual, infrared and electromagnetic imaging), depth (distance and resolution) and the quality of the data that is now available. With increasing worldwide availability of broadband data connections, the capability for rapid delivery of exceedingly large volumes of data continues to increase on an exponential basis.

The last part of this first criterion, the ability to create useful and actionable information based upon sensor data, is wholly limited by the imagination and innovation of researchers and the tools they create to detect objects and events relevant to seafarers and their missions. This includes all areas of engineering and deck operations and individual maritime tasks such as watchkeeping, engine operation, ship maintenance and navigation. The solution to much of this problem lies in part in success in automation of onboard systems and proper verification of development practices, validation of operational outcomes, and extensive testing to detect frailties of implementation.

#### 4.1.3.2 Situational Awareness and Comprehension

The comprehension of overall situational awareness to facilitate good, proper and correct decision making to ensure success in carrying out a safe voyage is ultimately the responsibility of captains and is based upon their own expertise and experience and in consideration of the thoughts and decisions of ship's executive officer, mates, staff and crew. For MASS, this translates into the capabilities of remote operators and onboard reasoning systems to correctly process the data and information provided by sensors and to assess their data quality, reliability and usefulness in performing decision making and solving problems. Further reliance is made upon the results and conclusions reached and actions taken by the individual shipboard automated processes to effect an overall desired outcome with respect to the voyage. Much of the solution to these issues lies in the realm of machine learning and artificial intelligence, subjects that were introduced in the previous chapter and are discussed in greater detail in the ensuing chapters.

#### 4.1.3.3 Sensor Suite Composition and Placement

Placement of the correct types of sensors at the proper locations on a ship is critical and unique to the requirements of the tasks being performed. For example, the use of position reference sensors combined with wind and motion sensors is essential to the task of dynamic positioning for ship's propellers and thrusters to maintain proper position and heading. However, these sensors are not suited to provide information on the amount of water that may exist and its displacement in the bilge, nor information on the depth of the water below the keel or the bottom terrain ahead of the vessel.

Many different types of sensors are placed throughout a vessel based upon their function and purpose, and no attempt will be made here to provide comprehensive treatment on the subject. However, among the most critical sensor applications for the successful completion of a voyage are those associated with sensing the environment, positioning and ship navigation as shown in Figure 4.1. This figure illustrates some of the types of shipboard sensors that provide situational awareness at and above the water's surface as well as below the waterline.

*Surface Sensors* – Located on fore and aft masts above the superstructure of the vessel can be Radar, Light Detection and Ranging (Lidar), millimeter Radar (mmRadar) and daylight, low-light imaging and infrared imaging systems that use different technologies to create increasingly high-resolution images and maps of the local environment to detect and recognize classes of fixed and moving objects and assist in collision and allision avoidance, and to identify specific objects that may be used as aids to navigation in determining position. The fusion of data from these different systems can complement each other to overcome the limitations of one or more systems under varying visibility and precipitation conditions.

The primary, long range Radar with Automatic Radar Plotting Aid (ARPA) provides a capability to detect both fixed and moving objects and to create tracks to calculate their course, speed and closest point of approach. Automatic Identification System (AIS) can provide similar capabilities. These are supplemented with short range, high-resolution Radar positioned as far forward as possible on the bow to detect close-in targets blocked by the bow that may be in the blind spot of the primary Radar as well as aids to navigation and hazards forward along the path of transit. Also residing on the fore and aft masts are Lidar systems that emit laser signals to create very high-resolution point-cloud imagery of the surrounding environment to detect, recognize and identify fixed and moving objects and predict their future motion for collision and allision avoidance. Production Lidar systems can provide an effective range from hundreds of meters to over one kilometer. mmRadar similar to those used in the Tesla automobile and military helicopters can also provide similar data products to create environmental maps for object detection, velocity determination and transit path projection. mmRadar is highly directional and it may be necessary to include multiple sensors fore, aft and along the sides of the vessel to provide coverage from all directions. Video camera imagery obtained through 360 degrees in the visible and infrared light spectrums supplement these previous technologies in their ability to create highly accurate environmental models with the added capability of heat detection that can be useful while performing search and rescue operations. Audio sensors (microphones) can be placed anywhere at unobstructed locations on the top of the hull and on the superstructure to listen for whistle signals from other vessels, horns activated from drawbridges, gongs and bells from buoys, fog horns and other audible signals useful for navigation.

MASS Design and Engineering 67

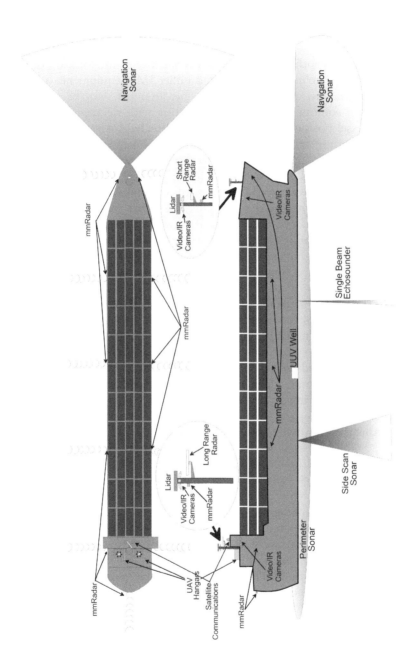

*Figure 4.1* MASS sensors and sensor locations for navigation.

*Subsea Sensors* – These are designed to enhance situational awareness below the waterline and consist of different types of devices that include single-beam echosounders, forward-looking navigation, side-scan and perimeter Sonars, each of which performs specific, specialized functions. The most common device is the single-beam echosounder (depth finder) that indicates the depth of the water below the keel at the location of the transducer. This instrument is mandated under IMO equipment carriage requirements and, in addition to depth measurement, can be used for bottom terrain contour following in correlation with electronic charts to determine vessel position. Navigation Sonar directed forward of the vessel along the path of transit provides high-resolution swath bathymetry to distances of 1,000 meters and beyond to determine bottom profiles in the area where the vessel is about to enter. This enables the detection of shoals, reefs, boulders, shipwrecks and other hazards to navigation that may exist in the water column such as shipping containers that are adrift and large marine mammals, and even buoys and the hulls of nearby vessels within close range whereby preventive and evasive action may be taken with sufficient time to avoid the hazard. Side-scan Sonar can create high-resolution imagery of large sections of the sea floor useful to detect natural and manmade features useful as aids to navigation in much the same manner as landmarks consisting of mountains and natural geological formations, towers and water tanks. Side-scan Sonar capabilities are presently being integrated with conventional single-beam echosounders by many manufacturers for fish-finding systems and are available at relatively low cost on the order of a few hundred dollars. Finally, perimeter Sonar, otherwise known as diver protection Sonar, is capable of detecting objects within close proximity to the vessel perimeter that can approach from any direction.

As with surface sensors, these different Sonar systems provide layered coverage that can dramatically enhance situational awareness below the waterline, enabling vessels to avoid dangers and hazards as well as providing backup navigation capabilities in the event of Global Navigation Satellite System (GNSS) failure, denial of service attacks or spoofing that can render existing satellite-based navigation inoperable or, worse, deceptive such as to aid hijacking and piracy encounters. One additional subsea sensor system is the remote or autonomous undersea vehicle (RUV/AUV) that can be launched from a well within the hull of the ship to explore the area and objects in the immediate vicinity as well as inspecting the ship's hull, propellers and rudder for damage. Live monitoring of RUV/AUV activity may be accomplished via video and telemetry data stream by a remote operator and an onboard reasoning system, where the findings and results may be considered for decision-making purposes.

*Airborne and Space Sensors* – Situational awareness beyond the horizon may be enhanced with the launch of one or more remote or autonomous aerial vehicles (RAV/AAV) to survey routes and inspect objects and hazards existing beyond the range of the ship's surface sensors. Such vehicles can provide live, multispectral imaging (e.g., daylight and infrared) and telemetry data streams for decision making by a remote operator on onboard reasoning system. These vehicles may be housed in one or more small hangars located in an environmentally sheltered area on the upper deck of the superstructure where there are no obstructions along the vehicle approach or departure paths. Access to space-based sensors may also be accomplished via broadband satellite communications with antennae mounted at locations on or above the ship superstructure that are without obstruction to view the sky.

### 4.1.4 Maintenance

A critical element that directly affects the reliability of MASS has to do with the maintenance requirements and practices needed to prevent and overcome equipment failures while underway. The solution to such problems is likely to be approached from four directions. First will be the design, implementation and installation of technologies and devices onboard vessels that will be more reliable and less prone to failure. Second, the implementation of system redundancy can be accomplished for critical systems whereby the failure of one system can be compensated for through a seamless interchange with another possessing redundant same capabilities without interruption of the process being performed. Third, better methods for accomplishing predictive maintenance can be developed to detect and identify failures that are likely to occur before they are realized during actual operations, and scheduling maintenance on those systems prior to sailing. The fourth direction is the development of robotic devices that can perform specific maintenance functions in the absence of onboard maintainers through remote control or autonomous operation.

#### 4.1.4.1 Introduction of Highly Reliable Systems

One of the best ways to lessen the impact of equipment maintenance that limit vessel operations is to improve system reliability while simultaneously lessening the need for human performance of maintenance tasks. New and improved technologies to replace older, less reliable systems and devices will go far towards improving overall reliability of MASS and enable their widespread use across many different application areas. One example is the case of high maintenance internal combustion diesel engines connected to complex drive trains that drive propellers. The transition from conventional low-RPM diesel propulsion driving a propeller, with a second diesel engine driving an alternator to handle other loads, to higher-RPM diesel-electric propulsion driving electric motors attached to the propeller and all other loads will lead to reduced operating costs; increased efficiency; improved reliability through redundancy in electricity generation capability; and decreased vibration, wear and tear on propulsion system components. An approach using banks of maintenance-free batteries capable of driving highly reliable electric motors connected directly to propellers, eliminating the diesel engine altogether along short routes, is inherently more reliable due to the significant reduction in moving parts for the propulsion system overall and the elimination of lubrication requirements for these parts.

Similar analogies can be made for significant improvements in overall reliability of a wide variety of vessel systems. Battery technology has improved from lead-acid devices that required periodic addition of fluids to low maintenance and high power sealed batteries using nickel-cadmium, lead-calcium, lithium-ion and other materials. Shipboard Radars, radios, Sonars and other electronic systems transitioned from vacuum tubes to semiconductors in the middle of the 20th century and greatly improved the reliability of electronic devices. The ongoing replacement of incandescent lighting with light-emitting diode (LED) technology on today's vessels provide yet another example for increased reliability with greater energy efficiency. Continuing improvements across all ship's systems in the technologies used in the development of sensors, computers, winches, electrical motors and other shipboard devices will result in overall improvement in terms of reduced maintenance tasks and costs and increased vessel reliability.

### *4.1.4.2 Multiple Redundant Systems*

The use of multiple, redundant systems to implement voting schemes as a means to improve complex system reliability have been accomplished for decades. Triple-modular redundancy was discussed as one method for improving computer reliability, variations of which have been deployed in aircraft and spacecraft with great success [Lyons and Vanderkulk 1962]. Physically redundant critical systems residing on MASS consisting of multiple assets of the same type and capabilities, where failed equipment may automatically be taken off-line and interchanged with properly operating equipment, can virtually eliminate single point of failure conditions that may render a vessel inoperable and prevent successful transition to a safe state. The need to successfully develop supporting architectures for this capability varies with each type of system, with greatly differing requirements posed by propulsion systems when compared to navigation and loading/unloading systems. In many cases, new communications bus protocols will need to be developed, or existing protocols modified and extended, which can facilitate automatic reconfiguration of system elements that goes beyond the "plug and play" (PNP) installation of these devices. One of many possible examples may be given for navigation instruments could include combining the capabilities of existing NMEA 0183/2000/future standards with the PCI eXtensions for Instrumentation (PXI) standard used for the rapid exchange of test instrument data, and extending the new standard to accomplish automated system reconfiguration. One of the benefits of this approach would be to include the modularization of all navigation instruments within one chassis with easily swappable, both physically and electrically, devices where maintenance and operation can be consolidated at one location. However, geographical distribution throughout the vessel of three chassis containing identical navigation instruments with independent electrical connections could provide the necessary redundancy to continue seamless operations in the event of failure or damage to one part of the ship due to collision, fire or other catastrophic event. Figure 4.2 illustrates an existing PXI system currently supporting electronic test equipment that could also host navigation instruments using a common chassis.

An embedded computer is shown on the left side of the chassis with test instruments immediately to the right of that. Several blank slots of varying sizes are illustrated from the center of the chassis to the far right of the chassis within which modular instruments hosting a combination single-beam and side-scan Sonar, AIS, GNSS, Radar, navigation Sonar and other instruments could be inserted.

Another method to further improve redundancy capability is through the employment of digital twin simulation where one or more redundant systems beyond those residing on MASS also exist at a remote control center to process the same onboard data and information communicated in real time through broadband networks. This would provide for immediate notification of shoreside or remote operators of problems as they occur, providing real-time recreation of the existing shipboard environment and offering opportunities for prompt attention by technicians to aid in their resolution. This approach would offer many of the same viewpoints in real time that are currently only available post incident through the recovery of a voyage data recorder.

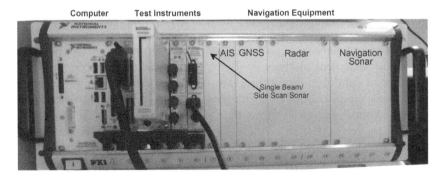

*Figure 4.2* PXI chassis supporting MASS test instruments and navigation system components.

### 4.1.4.3 Predictive Maintenance

The monitoring of system performance and nominal operating conditions of electrical, electromechanical and mechanical equipment using visual, infrared, electrical, vibration, electromagnetic, chemical and many other varieties of sensors can lead to the detection of conditions indicative of wear and pending failure before actual failure occurs. This is accomplished using a combination of statistical and artificial intelligence-based methods to proactively determine when individual system components will require maintenance.

Conditions and symptoms monitored can include:

- Vibrations resulting from lack of lubrication and wear in bearings and bushings,
- Increases in resistance and ringing in waveforms indicating pending electronic component failure,
- Trends in voltages and currents that are approaching upper or lower performance limits,
- Changes in speeds resulting from failing or dirty sensors in a feedback loop,
- Excessive heat, acoustic noise or electromagnetic interference originating in a system or component,
- Chemicals and contaminants in motor oils, hydraulic fluids, cooling water and other fluids, and
- Other conditions indicative of premature wear and/or pending failure.

These techniques are used to identify the precursors of failure and to predict potential anomalies that may lead to failure. Notification of anomalies and scheduling preventive maintenance is subsequently accomplished to avert pending failure and to establish estimates of remaining useful life for system components.

Predictive or conditional maintenance requires the constant acquisition and storage of large amounts of data from the many devices and equipment onboard MASS, including possibly some that may require retrofit of needed sensors into existing equipment to gather the necessary data. Historically it has been necessary to acquire the services and skills of experts and, with the aid of statistical analysis methods, to analyze and interpret the significance of condition monitoring data and the relationships between the data, and demonstrating their analysis through case studies associating eventual failure supported by the data. With the advent of machine learning and artificial intelligence these same data analytics can be performed by autonomous agents capable of spotting data trends and anomaly conditions as well as human experts and, in some cases, exceeding human capabilities where few human experts exist. Note that the joint hosting of test instruments and navigation equipment within the same chassis as shown in Figure 4.2 could lend itself towards a convenient architecture to achieve condition-based predictive maintenance for navigation and other onboard equipment.

One example of the benefits of predictive maintenance is provided by Caterpillar Marine, whose Asset Intelligence platform analysis of ship performance parameters quadrupled the cost to one customer for annual hull-cleaning expenses by decreasing cleaning intervals to every 6.2 months from once every two years, yet resulting in annual savings of up to $5 million and $400,000 per ship [Marr 2019]. Another example is the Royal Wagenborg offshore supply vessel *Kroonborg* test site for predictive maintenance for propulsion machinery where their condition-based SMArt Maintenance of Ships project is expected to result in significant cost reductions for the entire Wagenborg fleet [Wagenborg 2019].

### *4.1.4.4 Robotic Maintenance*

There exists great potential for the shipboard use of robotic technologies to perform significant tasks on MASS in terms of deck and superstructure maintenance, hull and ship inspection, as well as security in warding off hijackers [Kumar 2017; Ingham 2014]. The Naval Research Laboratory with U.S. universities that include Virginia Tech developed the Shipboard Autonomous Fire Fighting Robot (SAFFiR), an autonomous humanoid robot capable of detecting and suppressing shipboard fires. Originally intended for working with human firefighters, it can perform tasks that include turning valves, picking up and dragging fire hoses and putting water on a fire. Such capabilities can be adapted and put to effective use in performing many shipboard maintenance functions that require human dexterity and senses such as detecting and possibly repairing and/or replacing broken or damaged deck fixtures, cables, lashings, sensors and cameras along with many other problems. The visual, infrared and ultraviolet vision system combined with multimodal sensor technology for advanced navigation provides extraordinary vision even through smoke and fog and enables it to overcome obstacles and stay upright even in pitching and rolling sea conditions.

Robotic devices can also be used to examine and clean hull exteriors as well as examining interior tanks. The HullBUG has been developed by SeaRobotics and funded by U.S. Navy Office of Naval Research (ONR) to crawl on the hull surface and perform light cleaning of fouling films using a fluorometer to detect biofilm and rotary brushes or water-jets to scrub off the fouling film. Onboard sensors provide obstacle avoidance, path cleaning, and navigational capabilities. Another example is the RoboShip onrails robot that intelligently inspects a ship's tanks for potential damage developed by a team of Dutch and German companies and universities, including the University of Twente in the Netherlands. Similar products are also being developed by Alstom Inspection Robotics with their Ship Inspecting Robot (SIR) along with EU-funded projects to support the inspection of large marine vessels by using robotics systems such as the Marine Inspection Robotic Assistant System (MINOAS) and Inspection Capabilities for Enhanced Ship Safety (INCAAS).

The Recon Scout throwbot developed by Recon Robotics can be launched from secure deck locations at distances of up to 36 meters (120 ft) to crawl along a vessel and view piracy activities even in darkness using infrared imagery. This robot can also break into smaller bots to provide real-time imagery to remote controllers and onboard reasoning systems.

## 4.2 IMPLEMENTATIONS OF MASS

A brief look at some of the circumstances surrounding the remotely operated and autonomous vessels envisioned as well as in actual use today is provided in these paragraphs.

### 4.2.1 Container and Bulk Shipping

Short sea shipping of containers and bulk materials are early applications envisioned for MASS in harmony with the promise of automated port loading and unloading operations.

#### 4.2.1.1 Yara Birkeland

The first commercial implementation of MASS for container shipping is *Yara Birkeland*, a fully electric and zero emissions at point-of-use container feeder vessel [Skredderberget 2018]. The vessel was designed by engineering company Marin Teknikk (Gurshøy) and built by Vard (Ålesund) at a cost of approximately NOK 250 million, with NOK 133.6 million in support from the Norwegian government enterprise ENOVA [Yara 2018]. The hull is being delivered from Vard Bralla in Romania with vessel delivery from Vard Brevik in Norway in early 2020. Technology development is being accomplished by Kongsberg and the vessel is owned and operated by agricultural products company Yara International ASA (Oslo), both of Norway. Diesel-powered truck haulage is expected to be reduced by 40,000 journeys a year by moving container transport between Yara's Porsgrunn production plant from land to sea through Herøya to Brevik and to Larvik [MAREX 2017].

This 120-TEU vessel will have an overall length of 79.5 meters (261 ft), 3,200 tons deadweight and draft of five meters (16 ft) and is expected to begin operations in 2020. All operations will be in Norwegian waters between Herøya to Brevik over a distance of approximately 7 nautical miles, and Herøya to Larvik over a distance of approximately 30 nautical miles, with both routes covered by the Norwegian Coastal Administration VTS system at Brevik. A service speed of six knots with a maximum speed of 10 knots will be provided by two electric azimuth propeller pods and two tunnel thrusters using batteries with up to 9 MWh capacity. The batteries will also provide ballast weight for the vessel, thereby eliminating the need for ballast tanks. Sensors supporting autonomous operations include Radar, a Lidar device, an imaging system and an infrared camera with communications provided using a maritime broadband radio and a virtual private network connection via Inmarsat. Berthing and unberthing will be accomplished using an automatic mooring system enabling these operations to be performed without human intervention or special equipment on the quay. Cargo loading and unloading will also be accomplished automatically using electric cranes and equipment. All ship operations will be handled through three control centers to ensure safety involving ship monitoring, decision support, exception handling and emergency response.

### 4.2.1.2 Project SeaShuttle

Samskip, Europe's largest multimodal operator, is the lead partner in the Seashuttle project that seeks to develop next-generation sustainable short-sea shipping and bring emissions-free, autonomous container ships to market that also operate at a profit [MINN 2018]. This project was awarded €6 million from the Norwegian government for the development of two all-electric ships to provide container service between Poland, Swedish west coast ports and the Oslo fjord. The vessels will draw on state-of-the-art hydrogen fuel cells for their propulsion power to make it cost competitive with existing solutions as part of an approach for cost-effective and scalable competition with truck-ferry options that feed into a pan-European distribution network. Other Seashuttle partners include logistics consultant FlowChange, technology group Kongsberg Maritime, hydrogen integrator HYON and Massterly, a Kongsberg Maritime/Wilhelmsen venture developing autonomous vessel solutions.

### 4.2.1.3 Great Intelligence

The bulk carrier *Great Intelligence* is China's pilot smart ship project [Blenkey 2017]. The vessel was built in 2017 and measures 25,561 gross tons, 38,797 tons deadweight (180m length ×32m breadth with 6.8m draft, 590ft × 104ft × 22ft) with a maximum observed speed of 11.6 knots and average of 8.5 knots [MarineTraffic 2019b]. Although not an autonomous or remotely controlled ship in its own right, its automated systems can provide the basis for establishing autonomous ships with the addition of reasoning capabilities to supervise and coordinate multiple automated processes. The vessel has received Lloyds Register cyber-enabled ship (CES) descriptive notes Cyber AL2 Safe (Navigation, Propulsion, Steering), Cyber Al2 Maintain

(M/E, A/E, Boiler, Shaft) and Cyber AL2 Perform (Energy Management). Functions included in its onboard smart system capabilities include management of health and energy efficiency; integration of data management, analysis and forecasting; and intelligent navigation for route optimization and destination in shortest time, as well as most economical voyage.

### 4.2.2 Ferries

One of the most recent and successful maritime sectors where remote and autonomous operation has been demonstrated is with ferries, as illustrated by the following examples.

#### 4.2.2.1 Folgefonn

The passenger ship *Folgefonn*, owned by Norwegian ferry operator Norled, was operated autonomously in November 2018. Wärtsilä demonstrated full dock-to-dock autonomous operation for the entire route, visiting all three ports serviced by the ship [MAREX 2018]. The vessel measures at 1,182 gross tons, 597 tons deadweight (85m length ×15m breadth and 3.5m draft, 279ft × 49ft × 12ft) with a maximum observed speed of 13.3 knots and average of 9.8 knots [MarineTraffic 2019c]. The ferry demonstrated the capability to autonomously leave the dock, maneuver out of the harbor, sail to the next port of call, maneuver through the harbor entrance and dock alongside the terminal without requiring any human intervention. Vessel navigation was accomplished using a series of tracks and waypoints to guide the ship to the next destination with an autonomous controller system managing the vessel's speed, position on the pre-defined track and heading. GNSS was used as the primary sensor, while a Wärtsilä Guidance Marine CyScan system was tested as a secondary position sensor for the approach to the berth.

#### 4.2.2.2 Falco

In December 2018, Rolls-Royce and Finnish state-owned ferry operator Finferries performed a successful demonstration of the car ferry *Falco* in fully autonomous mode in the archipelago south of the city of Turku, Finland [Rolls-Royce 2018]. This vessel was built in 1993 and measures at 517 gross tons, 129 tons deadweight (52m length × 12m breadth and 4.2m draft, 170ft × 39ft × 14ft) with a maximum observed speed of 10.4 knots and average of 9.3 knots [MarineTraffic 2019d]. A combination of Rolls-Royce ship intelligence technologies was used to successfully navigate autonomously during its voyage between Parainen and Nauvo in Norway, while the return journey was accomplished under remote control. The demonstration was conducted with 80 invited guests aboard while the vessel detected objects using sensor fusion and artificial intelligence to perform collision avoidance and automatically alter course and speed when approaching the quay to carry out automatic docking without intervention from the crew. The vessel was equipped with a range of sensors that allow it to build a detailed, real-time picture of its surroundings with high accuracy that was relayed to Finferries' remote operating center on land 50 kilometers away in Turku city center, where a captain monitored the autonomous operations and could take control of the vessel if necessary.

### 4.2.2.3 Suomenlinna II

Asea Brown Boveri (ABB) and Helsinki City Transport in November 2018 were successful in remotely piloting the ice-class passenger ferry *Suomenlinna II* through a test area in Helsinki harbor, Finland [ABB 2018]. The vessel was built in 2004 and measures at 329 gross tons, 72 tons deadweight (33.8m length × 8m breadth with 3.9m draft, 111ft × 26ft × 13ft), with a maximum observed speed of 8.7 knots and average of 8.3 knots [MarineTraffic 2019e]. It is fitted with ABB's icebreaking Azipod® electric propulsion system and retrofitted with the ABB Ability™ Marine Pilot Vision situational awareness solution, the ABB Ability™ Marine Pilot Control dynamic positioning system, and steered from a control center in Helsinki. For the remote piloting trial, the ferry departed from Helsinki's market square, Kauppatori, and was wirelessly operated through a preselected area of the harbor.

## 4.2.3 Surveillance, Firefighting, Survey, and Search and Rescue

There are great potential markets for autonomous vessels in smaller lengths (up to 100 meters) for law enforcement, firefighting, and other commercial and military applications. Three of these initiatives are described below.

### 4.2.3.1 Sharktech

An example of an alliance to build small remotely operated and autonomous vessels is U.S. shipbuilder Metal Shark and U.K. autonomous vessel developer ASV Global with the development of the Sharktech autonomous vessel that will range in length from 5 to 100 meters across their full line of vessels using aluminum, steel and composite materials [Wingrove 2018; Metal Shark 2019]. These vessels are designed for use in remote and hostile environments, long endurance missions where loitering for long periods of time may be required, and for other missions where the use of human crews may not be warranted. They may also be configured for remote control, reduced manned and conventional operations according to mission requirements.

Sensor systems planned to support autonomous operations include the use of Radars, 360-degree daylight and infrared cameras and Automatic

Identification System (AIS) information to help avoid collision and allision. Navigation by waypoints is also featured [Metal Shark 2019]. Equipment capable of being integrated and operated by either autonomous or remote control includes firefighting, hydrographic survey, meteorological and oceanographic sensing, as well as specialty acoustic and camera devices.

#### 4.2.3.2 Sea Machines

A partnership between Sea Machines Robotics (Boston, USA) and workboat manufacturer Hike Metal (Ontario, Canada) will integrate Sea Machines' SM300 autonomous vessel control system aboard commercial vessels tasked with search and rescue (SAR) missions [Sea Machines 2019]. The SM300 system will add new capabilities to Hike Metal's vessels for autonomous search and rescue, data-driven waypoint following and mission planning, collaborative vessel operations, remote vessel and payload control, minimally manned and unmanned configurations, and obstacle avoidance. Demonstrations begin in summer 2019 near Hike Metal's headquarters on Lake Erie aboard a new-build, 27-foot rigid hull inflatable boat (RHIB) outfitted with twin 200-HP engines, a remote-controlled forward-looking infrared (FLIR) camera system, a heated survivor seating area, remote-controlled rescue cradles and extended-range fuel reserves.

#### 4.2.3.3 C-Worker 7

L3 Technologies in the United Kingdom has developed their *C-Worker 7* multi-role autonomous surface vessel capable of performing a wide variety of tasks and modular, interchangeable payloads for subsea positioning, surveying and environmental monitoring [L3 2019]. The vessel features an aluminum hull that is self-righting with a 2.5m × 1m moon pool, measuring at 5,300 kg deadweight (7.2m length × 2.3m breadth with 0.9m draft, 23.5ft × 7.5ft × 3ft), with 2 × 20kw of power and a maximum speed of 6.5 knots. It can be outfitted with a wide variety of sensors and communications is accomplished via radio and satellite.

### 4.2.4 Offshore Support

Offshore support vessels perform many and diverse functions for a variety of industries. A sampling of a few of these vessels that fall under the category of MASS is described in this paragraph.

#### 4.2.4.1 Hrönn

Automated Ships Ltd in the United Kingdom commissioned the building of *Hrönn* as the world's first autonomous, fully automated and cost-efficient prototype vessel for offshore operations in collaboration with the project's primary technology partner, Kongsberg [Kongsberg 2016]. At the time of this writing, it appears that plans for the construction of *Hrönn* have been cancelled.

#### 4.2.4.2 SeaZip 3

Autonomous trials of *SeaZip 3*, an offshore supply ship and crew transfer vessel (SeaZip NL), were held in March 2019 on the North Sea near Den Helder, Netherlands [Russell 2019]. The vessel was outfitted with sensors and software to facilitate several different scenarios for collision avoidance. The vessel was built in 2015 and measures at 168 gross tons, 20 tons deadweight (25.75m length × 10.1m breadth with 2.2m draft, 84ft × 33ft × 7.3ft) with a maximum observed speed of 21.3 knots and average of 20.1 knots [MarineTraffic 2019f]. This is a joint industry project supported by a number of shipping companies, suppliers and universities and is partly funded by the TKI-Maritiem allowance of the Dutch Ministry of Economic Affairs and Climate Policy.

#### 4.2.4.3 Autonomous Spaceport Drone Ships (SpaceX)

Probably the most famous examples of MASS are the Autonomous Spaceport Drone Ships (ASDS) presently in service to provide SpaceX launch vehicle recovery operations in the U.S. and elsewhere. Two vessels are presently operational and are named *Just Read the Instructions* and *Of Course I Still Love You* [Wall 2016]. A third vessel still under construction is named *A Shortfall of Gravitas* [Ralph 2018]. These vessels, chartered by SpaceX from McDonough Marine Service, are essentially their Marmac 300 ocean-going barges designed to maneuver autonomously using four azimuth thrusters made by Thrustmaster that allow for high-precision steering without a rudder [Prof. Mariner 2015]. They each measure approximately at 4,422 gross tons, 1,326 tons deadweight (91.4m length × 30.5m breadth with 6.0m draft, 300ft × 100ft × 19.9ft) [McDonough 2015].

### 4.2.5 Tugboat

The tugboat sector is also the focus of remote and autonomous operations as illustrated below.

#### 4.2.5.1 RAmora 2400

The uncrewed *RAmora* tug is the creation of naval architects Robert Allan Ltd (Vancouver, Canada) as part of their TOWBoT (tele-Operated Workboat or Tug) series of vessels designed to work in tandem with a conventional tug during ship handling [den Hertog, et al. 2016]. Featuring a hybrid propulsion system the vessel is fitted with Voith Schneider propeller drives in a fore/aft configuration. With substantial battery storage capacity, it is highly maneuverable and can provide a bollard pull of up to 55 tons in extended operations and even in hazardous conditions [RAL 2015]. The use of immersive telepresence features provides live 360-degree video and real-time electronic position-sensing to capture a continuous onboard perspective for safe and effective ship handling through remote operation from the command tug. A real-time control system provides the interface for the operator in addition to onboard maneuvering and positioning controls, equipment and workspace monitoring and safety management functionality.

### *4.2.5.2 Svitzer Hermod*

In 2017, Rolls-Royce and Svitzer demonstrated remote operation of the 28 meter (92 foot) tugboat *Svitzer Hermod*, another Robert Allan design, in Copenhagen harbor, Denmark including quayside berthing, turning through 360 degrees, and piloting back to Svitzer headquarters [Rolls-Royce 2017]. Built in 2016, the vessel measures at 461 gross tons, 132 tons deadweight (28.2 length × 12.6m breadth with 5.5m draft, 93ft × 41ft × 18ft), with a maximum observed speed of 10.5 knots and average of 8.3 knots [MarineTraffic 2019g]. It uses a Rolls-Royce dynamic positioning system that facilitates remote control. Vessel propulsion is accomplished using two Rolls-Royce MTU 16V4000 M63 diesel engines rated 2000 kW at 1800 rpm. Onboard sensor data communicated to the remote operating center located at Svitzer headquarters at the harbor provided the captain with situational awareness of the vessel and its surroundings [Am. Shipper 2017]. Also demonstrated was a concept for a future proof standard for remotely controlled vessels through the optimal placement of the different system components to give the master high levels of confidence and control.

### *4.2.5.3 Keppel Singmarine*

Specialized shipbuilder Keppel Singmarine Pte Ltd is developing an autonomous tug, expected to be one of Singapore's first autonomous vessels, scheduled to be completed in late 2020 and operated by Keppel Smit Towage [Offshore 2019]. This project involves the modification of a 65-metric-ton tug by retrofitting the vessel with position maneuvering, digital pilot vision, collision detection and avoidance systems as well as establishing an onshore command center to remotely control the tug. Communications will be accomplished through a Keppel Singmarine partnership with Keppel Group M1 to leverage its ultra-low latency 4.5G network connectivity to ensure high reliability for ship-to-shore communication and to support Internet of Things maritime applications. Keppel Offshore & Marine, the Singapore Maritime Port Authority (MPA) and the Technology Centre for Offshore and Marine Singapore (TCOMS) are working together under a memorandum of understanding signed in April 2018 to jointly develop autonomous vessels for a variety of applications including undertaking harbor operations such as channeling, berthing, mooring, and towing operations.

## 4.3 HARBOR ENHANCEMENTS TO ACCOMMODATE MASS

The continuing improvements exhibited in port infrastructure through automation and new technology to aid in vessel handling process enhancement has led to increased efficiencies and improved safety in many areas. There are several areas where improvements may still be made that will be beneficial to shipping in general, but will specifically benefit MASS in terms of achieving autonomy. These include the port facilities, the harbors themselves and the approaches to the harbors where the most congested waters are most likely to be experienced.

### 4.3.1 Port and Harbor Facilities

Several different aspects of port operations are presently experiencing improvements that will enhance the safety of seafarers as well as improving efficiencies in port operations. These have to do primarily with docking and bunkering. Automated loading and unloading of cargo from MASS should be relatively similar to conventionally manned vessels.

#### 4.3.1.1 Automatic Berthing and Unberthing

The docking process has long been a hazardous undertaking with the ever-present danger of snapback from one or more lines and wire ropes that may part without warning to injure, maim and even kill seafarers on deck and shore personnel quayside. To address this issue the use of new technology has been considered to eliminate lines and ropes altogether in securing a vessel to the wharf. As early as 2003 the port of Rotterdam in the Netherlands was experimenting with the use of a series of strong electromagnets built into the quay to moor giant container ships [Keulemans 2003]. The concept involved the use of a series of electromagnets each of which generates a 1-tesla magnetic field that is concentrated near the vessel such that minimal interference with shipboard devices was anticipated. It was estimated that 52 of these magnets mounted along a quayside will be capable of holding a 400-meter container ship in place unaffected by the wash from passing ships and in winds of up to storm force 12. An alternative system was also being developed during the same time by Mooring Systems in Christchurch, New Zealand, using shore-mounted vacuum pods that cling to the side of a vessel. This system was almost entirely mechanical, using electric power only to attach the pods to the vessel.

Today these systems appear in several forms that are usable by MASS. Mampaey Offshore Industries (the Netherlands) describes their DockLock© automatic magnetic mooring system as a safe approach that eliminates injury risk with less exposure time and provides live monitoring of mooring operations as well as external influences and conditions that can affect docking [van Reenen 2019]. This unique design can secure a ship in less than 1 minute and decouple a ship in less than 20 seconds, providing faster turnaround time, better ship utilization and faster response time to emergency situations. Cavotec (Switzerland) provides similar functionality with their MoorMaster™ automated mooring system that uses vacuum pads to more and release vessels [Cavotec 2019]. The system is capable of mooring a vessel in less than 30 seconds and provides reduced vessel motion from swell, surge and passing ships. Any vessel can use the berth equipped with the system, even vessels that are longer than the berth. Both systems result in reduced emissions resulting from decreased use of tugs and ship engines.

Raymarine has also introduced their DockSense™-assisted docking technology that uses machine vision object recognition with FLIR imagery from five cameras mounted on the vessel and motion sensing integrated with vessel propulsion and steering to assist in tight quarter docking maneuvers and for close-in collision avoidance [Rudow 2019]. The system uses GNSS and an attitude heading reference system to effectively implement an electronic bumper stopping the vessel to avoid contact with the quay, pilings and other vessels.

### *4.3.1.2 Bunkering*

One area in which automation is lagging is bunkering. The most modern of automated ports supporting MASS are still likely to involve manual tasks to connect between the vessel and bunkers source. This may be due in part to the historically conservative nature of the shipping industry, but more often than not this is a broad generalization since many in the shipping industry are quick to embrace change. Rather, the source of this perceived reluctance may stem from potential liability associated with environmental damage and cleanup in the event of mishap during bunkering.

One bright spot in this topic is in the area of recharging ships' batteries for electrically powered vessels. This is accomplished using inductive wireless power transfer for safe and fully automated operations with better utilization of the docking time for charging the batteries [Guidi, et al. 2017]. An example of power transfer in the range of 1–2 MW is that being developed for the plug-in zero emission hybrid Norwegian ferry *Folgefonn* discussed earlier in this chapter.

EST-Floattech (the Netherlands) is collaborating with Integrated Infrastructure Solutions (Germany) to develop an Inductive Charging System® for the marine sector [EST-Floattech 2017]. This system will be combined with the lithium-ion battery chemistry and is specifically designed for electric and hybrid power propulsion and is intended for application on yachts, commercial vessels and ferries. It eliminates the need for cable connections and also reduces the need for maintenance as there are no cables that can wear out or electrical connection points to become damaged or worn by water, seawater, snow and ice.

### 4.3.2 Harbor Approaches

Conventional aids to navigation and buoyage within harbors and their approaches are designed for human vision (colors, markings and lights), hearing (gongs, bells, horns and whistles) and Radar reflectivity for vessels to safely navigate within the confines of channels and among other vessels. Existing IMO and national regulations presently do not recognize machine vision as a viable watch standing tool in the absence of seafarers. China, Singapore, the Netherlands and the Scandinavian nations are operating MASS within special testing areas to get around legal and regulatory considerations, while the United Kingdom is treating all territorial and inland waters as open for MASS to operate under license from the appropriate, authorized statutory authorities [ShipTech 2018]. Presently, few provisions exist for test bed locations for MASS research and development within the United States.

There is an opportunity to create entire classes of alternative and new designs for aids to navigation that are optimized for MASS sensor requirements. One such system is represented by virtual aids to navigation that require no physical infrastructure such as an AIS transmitter to properly watch navigable waters that is described in Chapter 6, "Navigation". This may also include Sonar systems that can be adapted to navigate using bottom contours along with natural and manmade features of the undersea environment. New technologies are likely to be developed and introduced within the next decade to accomplish these goals.

### REFERENCES

ABB 2018. ABB Enables Groundbreaking Trial of Remotely Operated Passenger Ferry. Group Press Release, Asea Brown Boveri Ltd, Zurich, Switzerland, 4 December 2018. https://new.abb.com/news/detail/11632/abb-enables-groundbreaking-trial-of-remotely-operated-passenger-ferry.

American Shipper 2017. Rolls-Royce, Svitzer Demo Remotely Operated Tug Tech. American Shipper, 21 June 2017. https://www.americanshipper.com/news/?autonumber=67881&source=redirected-from-old-site-link.

Ash, Nick and Scarbrough, Tim. 2019. *Sailing on Solar: Could Green Ammonia Decarbonize International Shipping?* Environmental Defense Fund Europe Ltd., London, 2019.

Blenkey, Nick. 2017. Chinese Smart Ship Earns LR Cyber Notations. *MarineLog*, 8 December 2017. https://www.marinelog.com/news/chinese-smart-ship-earns-lr-cyber-notations/.

Brandslet, Steinar 2019. Giving Ferries Wings to Optimize Wave Power. *Noprwegian SciTech News*, 2 July 2019. https://norwegianscitechnews.com/2019/07/giving-ferries-wings-to-optimize-wave-power/.

Cavotec 2019. Automated Mooring. Cavotec SA, Lugano, Switzerland. www.cavotec.com/en/your-applications/ports-maritime/automated-mooring.

den Hertog, Vinve, Oscar Lisagor, Robin Stapleton and Chris Kaminski. 2016. Revolutionary RAmora Brings Tele-operated Capability to Ship-Handling Tugs. International Tug, Salvage and OSV Convention, Boston, MA. Day 2, Paper 4.

Eidesvik 2019. Eidesvik Offshore ASA. Bølmo, Norway. https://eidesvik.no/supply/viking-energy-article262-868.html.

Equinor 2018. *This Electric "Viking Ship" Turned Out Even Better than Expected*. Equinor ASA, Stavanger, Norway, 2018. https://www.equinor.com/en/magazine/battery-hybrid-supply-ship.html.

EST-Floattech 2017. Inductive Charging System, Press Release. EST-Floattech, The Netherlands, 17 February 2017. https://www.est-floattech.com/inductive-charging-system/.

FCHEA 2018. Shipping Propulsion. Fuel Cell and Hydrogen Energy Association (FCHEA), Washington DC, October 22, 2018. http://www.fchea.org/in-transition/2018/10/23/shipping-propulsion.

GMFT 2030. Global Marine Fuel Trends. 2030. Lloyds Register Marine, pp. 31–7.

Gronholt-Pedersen, Jacob. 2018. Maersk Tankers Tests Wind Power to Fuel Ships. Reuters, 30 August 2018. https://www.reuters.com/article/us-shipping-fuel-windpower/maersk-tankers-tests-wind-power-to-fuel-ships-idUSKCN-1LF1WW.

Guidi, Giuseppe, Jon Are Suul, Frode Jenset and Ingve Sorfonn. 2017. Wireless Charging for Ships: High-Power Inductive Charging for Battery Electric and Plug-In Hybrid Vessels. *IEEE Electrification Magazine*, 5, Issue 3: 22–32. DOI:10.1109/MELE.2017.2718829.

IMO PAPS. Regulations for the Prevention of Air Pollution from Ships (Annex VI). MARPOL Convention, Intermational Maritime Organization, 2020 Sulphur Limit FAQ 2019.pdf.

Ingham, Lucy. 2014. Autonomous Robotic Ship Inspector to Cut Costs and Boost Safety. *Factor Magazine*, 21 November 2014. https://www.factor-tech.com/robotics/9735-autonomous-robotic-ship-inspector-to-cut-costs-and-boost-safety/.

Keulemans, Maarten. 2003. Giant Electromagnets to Moor Ships. *New Scientist*, 17 January 2003. https://www.newscientist.com/article/dn3270-giant-electromagnets-to-moor-ships/.

Kongsberg 2016. Automated Ships Ltd and Kongsberg to Build First Unmanned and Fully Autonomous Ship for Offshore Operations. 5 November 2016. https://www.kongsberg.com/maritime/about-us/news-and-media/news-archive/2016/automated-ships-ltd-and-kongsberg-to-build-first-unmanned-and-fully-autonomous/

Kumar, Sukant 2017. 5 Innovative Robotic Technologies for the Maritime Industry. Marine Insight, Future Shipping, 13 September 2017. https://www.marineinsight.com/future-shipping/5-innovative-robotic-technologies-for-the-maritime-industry/.

L3 2019. L3 Technologies. C-Worker 7 Work Class ASV. Technical Specifications, 2019.

Lyons, R. E. and W. Vanderkulk. 1962. The Use of Triple-Modular Redundancy to Improve Computer Reliability. *IBM Journal*, April 1962, pp. 200–9.

MAREX 2017. Yara Birkeland Model Testing Underway. *Maritime Executive*, 9 October 2017. https://maritime-executive.com/article/yara-birkeland-model-testing-underway#gs.Ctlrtaw.

MAREX 2018. Wärtsilä Conducts Autonomous Ferry Voyage and Docking. *Maritime Executive*, 28 November 2018. https://www.maritime-executive.com/article/waertsilae-conducts-autonomous-ferry-voyage-and-docking.

MAREX 2019. Vertical Turbine Developed for Onboard Renewable Energy. *Maritime Executive*, 22 May 2019. https://maritime-executive.com/article/vertical-turbine-developed-for-onboard-renewable-energy.

MarineTraffic 2019a. *Viking Energy*. https://www.marinetraffic.com/en/ais/details/ships/shipid:311799/mmsi:258390000/imo:9258442/vessel:VIKING_ENERGY.

MarineTraffic 2019b. *Great Intelligence*. https://www.marinetraffic.com/en/ais/details/ships/shipid:5320084/mmsi:477151900/imo:9800623/vessel:GREAT_INTELLIGENCE.

MarineTraffic 2019c. *Folgefonn*. https://www.marinetraffic.com/en/ais/details/ships/shipid:314185/mmsi: 259530000/imo:9172090/vessel:FOLGEFONN.

MarineTraffic 2019d. *Falco*. https://www.marinetraffic.com/en/ais/details/ships/shipid:6603/mmsi:230987390 /imo:8685741/vessel:FALCO.

MarineTraffic 2019e. *Suomenlinna II*. https://www.marinetraffic.com/en/ais/details/ships/shipid:6552/mmsi:230985490/imo:9315408/vessel:SUOMENLINNA_II.

MarineTraffic 2019f. *SeaZip 3*. https://www.marinetraffic.com/en/ais/details/ships/shipid:3372516/mmsi: 244830667/imo:9758686/vessel:SEAZIP_3.

MarineTraffic, 2019g. *Svitzer Hermon*. https://www.marinetraffic.com/en/ais/details/ships/shipid:4655209/mmsi:219022265/imo:9788124/vessel:SVITZER_HERMOD.

McDonough 2015. SpaceX charters McDonough Marine barge. Professional Mariner. 3 March 2015. www.professionalmariner.comWeb-Bulletin-2015/SpaceX-charters-McDonough -Marine-Barge.

Metal Shark 2019. Introducing Sharktech Autonomous Vessels. Metal Shark, May 2019. www.metalshark.com/autonomous-vessels/.

MINN 2019. MI News Network. Samskip Takes Lead In Initiative To Develop Autonomous, Zero-Emissions Container Ships. Marine Insight. December 24, 2018. https://www.marineinsight.com/shipping-news/samskip-takes-lead-in-initiative-to-develop-autonomous-zero-emissions-container-ships/.

Offshore 2019. Offshore Staff. Keppel Developing Autonomous Tugboat, 10 April 2019. https://www.offshore-mag.com/rigs-vessels/article/16790823/keppel-developing-autonomous-tugboat.

Oh, Se-Young. 2018. Low Carbon Technology for Marine Application. Hyundai Heavy Industries, 29 November 2018. https://lngfutures.edu.au/wp-content/uploads/2018/12/9-Hyundai-Heavy-Industries-Hydrogen-workshop HHI-Final.pdf.

RAL 2015. Revolutionary REmora Brings Tele-operated Capability to Ship Handling. Robert Allan, Ltd, 18 September 2015. https://ral.ca/2015/09/18/revolutionary-ramora-brings-tele-operated-capability-to-ship-handling/.

Ralph, Eric. 2018. The Vessel Is to be Named *A Shortfall of Gravitas*, 12 February 2018. www.teslarati.com/spacex-new-drone-ship-a-shortfall-of-gravitas.

Rolls-Royce 2017. Rolls-Royce Demonstrates World's First Remotely Operated Commercial Vessel. Rolls-Royce, Press Release, 20 June 2017. https://www.rolls-royce.com/media/press-releases/2017/20-06-2017-rr-demonstrates-worlds-first-remotely-operated-commercial-vessel.aspx.

Rolls-Royce 2018. Rolls-Royce and Finferries Demonstrate World's First Fully Autonomous Ferry. Rolls-Royce, Press Release, 3 December 2018. https://www.rolls-royce.com/media/press-releases/2018/03-12-2018-rr-and-finferries-demonstrate-worlds-first-fully-autonomous-ferry.aspx.

Rudow, Lenny 2019. New Releases. Electronics. *Boat US Magazine*, June–July 2019, p. 44.

Russell, Tom. 2019. Autonomous Manoeuvring Trials Comleted. C4 Offshore, 27 March 2019. https://www.4coffshore.com/news/autonomous-manoeuvring-vessel-trials-concluded-nid12436.html.

Sea Machines 2019. Sea Machines Partners with Hike Metal to Demonstrate Capabilities of Marine Autonomy during Search-and-Rescue Missions. Sea Machines, May 14, 2019. https://sea-machines.com/sea-machines-partners-with-hike-metal-to-demonstrate-capabilities-of-marine-autonomy-during-search-and-rescue-missions/.

ShipTech 2018. How Should Ports Prepare for Autonomous Shipping? Ship Technology, 3 December 2018. https://www.ship-technology.com/features/ports-autonomous-shipping/.

Skredderberget, Asle 2018. The First Ever Zero Emission, Autonomous Ship. Yara International ASA, Olso, Norway, 14 March 2018. https://www.yara.com/knowledge-grows/game-changer-for-the-environment/.

Ulstein 2019. Color Hybrid. https://ulstein.com/references/color-hybrid.

van Reenen, Wouter 2019. Automatic Magnetic Mooring. Mampaey Offshore Industries, 2019. https://www.unece.org/fileadmin/DAM/trans/doc/2013/dgwp15ac2/Dock_lock_presentation_1.pdf.

Wagenborg 2018. Cost Reduction through Preventive Maintenance. Royal Wagenborg, Delfzijl, The Netherlands, 4 March 2018. www.wagenborg.com/cases/cost-reduction-through-preventive-maintenance.

Wall, Mike. 2016. Elon Musk Names SpaceX Drone Ships in Honor of Sci-Fi Legend. space.com, 4 February 2016. www.space.com/28445-spacex-elon-musk-drone-ships-names.html.

Wingrove, Martyn. 2018. Alliance Unveils Autonomous Vessel Technology. Maritime Digitalisation & Communications, 16 July 2018. http://www.marinemec.com/news/view,alliance-unveils-autonomous-vessel-technology_53510.htm.

Yara 2018. Yara Selects Norwegian Shipbuilder VARD for Zero-Emission Vessel Yara Birkeland. Corporate Releases, Brevik, 15 August 2018. https://www.yara.com/corporate-releases/yara-selects-norwegian-shipbuilder-vard-for-zero-emission-vessel-yara-birkeland/.

Chapter 5

# Remote Control Centers

Maritime Autonomous Surface Ships (MASS) will bring disruptive changes to ship design and management that requires new thinking with respect to the roles of Captains and their crews in vessel operation, new infrastructure necessary to connect ships to the shore and other remote locations, and the facilities to monitor and supervise these vessels. This includes the technologies required to analyze and act upon the enormous amounts of data inherent to this undertaking. The Remote Control Center (RCC) embodies the heart of MASS operations where, as on a ship, the ultimate authority exists to claim credit for a successful voyage and to assume responsibility for problems and the consequences of failure. The maritime culture has centuries of experience in formulating rules, customs and traditions to ensure safety of navigation, promote effective vessel transits and the discipline needed to accomplish these goals. This heritage should be considered as a valuable asset and not be minimized when transitioning this authority from the ship to the shore and other remote locations with the implementation of MASS. A portion of this authority will now be delegated to an automated reasoning system to control a vessel yet ultimate responsibility must be retained by the Captain and crew at the RCC and, indirectly, the reasoning system software developer and the builders of MASS.

From the beginning when ships first voyaged beyond the horizon, the vessel and her crew together formed a fully autonomous entity that was entirely responsible for the outcome of their voyage. Over the years the management of shipping companies has endeavored to maintain and improve contact with their ships and Captains. Current events merely represent a new approach towards solving the same types of problems encountered for centuries to gather data and to communicate between remote locations and their ships. MASS represents the ultimate in communications where the intent, instructions and experience of those giving the instructions can be conveyed and carried out by the ship itself.

Such communications have always taken advantage of the best available technology of their era beginning with documents carried by couriers. With the advent of radio, these evolved into coded messages of a few lines in length and voice transmissions to provide instructions and weather advisories, effect changes in routing, and inform of disruptions in the logistical chain and other events where action onboard the vessel had to be initiated remotely. Exchange of information from the ship to a remote location includes voyage status and vessel performance reports relayed by voice and in electronic form. The arrival of ubiquitous broadband communications in the early 2020s via a combination of terrestrial and satellite links provides the capability to transfer the unprecedented huge volumes of bi-directional data traffic that will make possible the remote operation and the direct and indirect supervision of autonomous vessels.

## 5.1 TRANSITION TO REMOTE CONTROL AND SUPERVISION

The modern practice of recording voyage data as a means to gain awareness into events that occur on vessels extends well into the last century. Initial reporting between ship and shore was accomplished using high-frequency (HF) radio by a trained radio operator using Morse Code and via telegraph messaging using Teleprinter exchange (TELEX). These modes were eventually supplemented with voice communications. The passage of regulations mandating the use of voyage data recorders (VDRs) on vessels beginning in 2002 for all passenger ships and other vessels of greater than 3,000 gross tons initiated the automatic recording of relatively large volumes (when compared to pre-VDR use) of vessel data regarding the operation and performance of critical shipboard systems.[SOLAS V] These data include date and time, position, speed and heading, bridge audio, radio communication audio, Radar, Electronic Chart Display Information System (ECDIS), echosounder, alarms, rudder order and response, hull opening status, watertight and fire door status, speed and acceleration, hull stresses, wind speed and direction recorded over a period of at least the previous 12 hours. Since ship-to-shore communications were highly restricted in terms of bandwidth, availability and reliability, vessel data was recorded and stored in these "black boxes" to be examined upon reaching port, or after recovery in the event of the loss of the vessel. Examples of beneficial knowledge gained using VDR data include revelations that under keel clearances had been ignored and propeller pitch controls were not used effectively [North 2015]. It also allowed for data from a vessel entering a new port to be passed to other vessels to assist in their voyage planning.

Beginning around 2002 with expanding cruise ship traffic and larger cargo vessels combined with advances in data analytics and increasing satellite bandwidth, the shipping industry expanded their efforts to find ways to improve vessel efficiency. One company, Eniram (Helsinki, Finland), in 2005 focused on gathering and analyzing performance data to optimize energy consumption onboard cruise vessels [MAREX 2016]. Their process gathered over a trillion measurement points onboard a vessel during normal operation to build a mathematical model of the relationships between different performance elements onboard and the relationship between the ship and the sea with all the affecting factors. Using statistical analysis methods, Eniram was successful in bringing real-time guidance to onboard engineers and deck officers on how to best trim the vessel, what speed to operate, how much load the engines should have and what route to take. At present the company optimizes the operations of more than 130 large cruise vessels, or about two-thirds of the world's seagoing fleet. There currently exist several commercially available products from various suppliers with not only data monitoring but maintenance and diagnostic capabilities for many ship systems.

Present-day MASS is taking this concept of monitoring vessel functions and performance by an operations and monitoring center on shore to new heights by exponentially increasing the volume and types of real-time data acquired through the capture of both telemetry and imagery from a vastly increased number of onboard sensors throughout a vessel. This includes sensors embedded within major systems for navigation, propulsion, power, alarms, vessel motion, the hull and superstructure and many of their deck components. Furthermore, the amount of data being transmitted in the opposite direction enabling remote control and operation of its functions also greatly increases the data volume that is communicated, which makes possible transitioning the shoreside operations monitoring center into a MASS RCC. The comprehensive analysis of these data through the use of statistical and artificial intelligence tools will result in unprecedented degrees of situational awareness and insight into vessel performance by both onboard reasoning systems residing on MASS and the ships' crews and shipping companies back on shore and at remote locations that may include other vessels. The need for comprehensive MASS RCC facilities has been identified and discussed for years and is swiftly progressing towards implementation for both civilian and military control of autonomous vessels [Rødseth 2014; Ottesen 2015, MUNIN 2016a; Roll-Royce 2016; Baldauf, et al. 2017; Aamaas 2018].

## 5.2 REMOTE CONTROL CENTER FUNCTIONS

The functions performed at the RCC vary widely with respect to the degree of autonomy at which MASS operate. Considering that one center can potentially be responsible for many MASS, an RCC should also anticipate multiple, simultaneous operations at various degrees of autonomy. These functions will vary depending on factors that include:

- Degree of human participation in ship operation,
- Levels of accessibility to the vessel for monitoring, command and control,

- Location of decision making ranging from full on board human control to remote operation from shore or another remote location and unsupervised autonomous operation,
- Scale of autonomy with scope from none through partial and full autonomy, and
- Extent of authority and capability to perform and intervene in ship operations.

In the Final Report on their Analysis of Regulatory Barriers to the use of Autonomous Ships submitted by Denmark, a proposed approach on the basis of four degrees of autonomy (as adapted to IMO autonomy designations) important from a regulatory perspective is illustrated in Table 5.1 [MSC 99/INF.3.]. These functions of the RCC directly translate into the requirements for and capabilities needed of the Captains and crews of these vessels, the facilities and their support staff to implement this approach, data and sensor analytics, and reasoning capabilities that provide insight into the significance of the vast amounts of information to assist the ships' crews in maintaining safety of navigation as well as provide the management of shipping companies with the current status of their operations. A discussion of these topics follows.

*Table 5.1* IMO Degrees of Autonomy Supported by a Remote Control Center

| | |
|---|---|
| **IMO Degree 1 – Ship with automated processes and decision support** | The operator (master) is on board controlling the ship that is manned as per current manning standards. Subject to sufficient technical support options and warning systems, the bridge may at times be unmanned with an officer on standby ready to take control and assume the navigational watch. |
| **IMO Degree 2 – Remotely controlled ship with seafarers on board** | The vessel is controlled and operated from shore or from another vessel, but a person trained for navigational watch and maneuvering of the ship will be on board on standby ready to receive control and assume the navigational watch, in which case the autonomy level shifts to Degree 1. |
| **IMO Degree 3 – Remotely controlled ship without seafarers on board** | The vessel is controlled from shore or from another vessel and does not have any crew on board. |
| **IMO Degree 4 – Fully autonomous ship** | The operating system of the vessel calculates consequences and risks. The system is able to make decisions and determine actions by itself. The operator on shore is only involved in decisions, if the system fails or prompts for human intervention, in which case the autonomy level will shift to Degree 2 or 3, depending on whether or not there is crew on board. |

Sources: Degrees of Autonomy: MSC 99-22. Descriptive text adapted to IMO MASS autonomy definitions (MSC 100) from: Analysis of Regulatory Barriers to the use of Autonomous Ships. [MSC 99/INF.3]

## 5.2.1 Distribution of Authority

The Captain holds ultimate authority and responsibility on a ship, and without his or her approval nothing can (or should) happen. In practice, much of this authority is delegated to the ship's departments that include engineering, deck, security and other traditional functions, where the senior officers report directly either to the Captain or to the First Mate/Executive Officer, depending on the command structure. The Captain is also responsible for conducting ship's business both at sea and in port, which includes overseeing cargos along with loading and unloading operations, meeting the requirements of local authorities, customs and inspections; compliance with environmental regulations, crew lists and assignments, financial operations and accounting, and handling issues pertaining to vessel accidents, incidents and losses; and the keeping of ship's documents, certificates and log books. Again, much of this authority is delegated to ship's departments as well as remote support personnel who are responsible for keeping the Captain informed of the status of ongoing operations and scheduling, problems and their resolution, and actions taken and needed in the future.

In the case of MASS, some of these lines of authority may tend to become blurred through the need for and the ready availability of additional staff that are not traditionally assigned duties at sea. This may also occur through attempts at multitasking Captains, ship's officers and crews by spreading their responsibilities across multiple vessels where full time attention to one vessel may not appear to be needed or warranted as a result of efficiencies gained through automation. Such issues are entirely new to the maritime industry in general and present unique dilemmas, challenges and opportunities for shipping to evolve into the future.

### 5.2.1.1 Individual Ship Operations

MASS integrate existing maritime job descriptions with completely new positions and responsibilities unique to remote and autonomous functions that must be tailored to the needs of maritime operations. Figure 5.1 illustrates a simplified summary of several of these jobs and functions as they pertain to the operation of one MASS entity from an RCC. This paragraph discusses one of several possible methods for implementing such an approach.

At the *Management Level* for conventional ships as well as MASS, the Captain must continue to remain the ultimate authority. This also means that the organizational structure of the senior officers and crew upon which he or she relies including the First Mate and Chief Engineer and bridge team and department heads must also remain intact. However, considering the unique nature of MASS and the additional features and equipment present on these vessels, the scope of the Captain's duties and responsibilities may need to be expanded to cover these new aspects of the job with additional crew members and support staff. Likewise, the same applies to

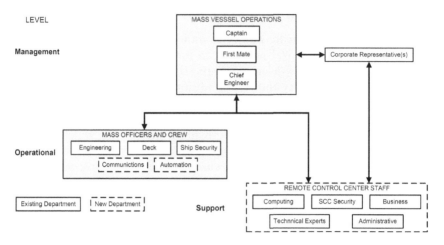

*Figure 5.1* Chain of authority for individual MASS operations.

senior officers, where there exists the potential for the addition of one or two more ship's departments (communications and automation) to reflect the unique requirements for MASS.

The *Operational Level* should continue to reflect tasks for seafarers serving as officer in charge of a navigational or engineering watch or as designated duty engineer. The nature of the tasks performed should remain relatively the same, while the methods used to perform these tasks may change significantly as new devices and methods are identified that take advantage of comprehensive sensor coverage throughout the ship and extending into the surrounding environment. The role of the Security Officer will change not so much in terms of the types and nature of threats likely to be encountered by an individual vessel but rather in terms of the methods and tools to be used to detect, repulse and overcome them. Cyber threats will continue to increase in significance as new methods are developed to target individual vessels and co-opt their functions and capabilities.

Changes are also needed at this level to reflect the addition of new members of the crew knowledgeable in MASS-specific communications beyond current Global Maritime Distress Safety System (GMDSS) requirements as well as those involved with the many new sensors that reside on MASS. This includes the reintroduction of the out-dated position of radio officer possibly as a Communications Officer who will oversee all aspects of satellite and terrestrial communications in terms of sensor data and telemetry originating from the vessel as well as command and control communications originating from the RCC. Also, with sensors providing the sole method for the RCC to gain situational awareness of the vessel, the environment in which it operates and all of its systems, the function of overseeing sensor operation and performance gains unprecedented significance and importance as compared to conventional vessels.

It is at the *Support Level* where crew members have traditionally been assigned to perform tasks on board a seagoing ship under the supervision of an individual serving in the Operational or Management Level. Historically this is the level where the most significant changes are likely to be found in the makeup of MASS crew members and support staff. Advances in technology in commercial shipping over the years have made obsolete the role of rigger to work the running rigging and to furl and release the sails; of oarsman to propel and maneuver vessels in close quarters; of pigeoneers to feed and care for pigeons delivering messages between ship and shore; of carpenters to repair ship's timbers and spars, coal heavers to haul coal from a ship's bunker to the boiler; and of the ship's cooper to make and repair barrels, casks and buckets at sea. The next likely candidate is the role of deck hand, as automated mooring and docking systems become more capable and widespread. Robotic systems are very likely to take over routine maintenance of a decreasing number of individual deck equipment installations at sea as MASS encompass more systems closed to the environment, while Able Seafarers transition more to standing navigation and security-related watches.

Many present-day Support Level tasks will migrate to the RCC while others will be eliminated entirely through automation. However, new jobs will also be created. New support departments are likely to be created around the functions of analyzing the communications and reasoning capabilities onboard MASS to ensure the vessel is performing in a predictable, reliable and repeatable manner. Also, while the Vessel Security Officer VSO is considering the various threats and activities related to a single vessel or group of vessels to which he or she is assigned, the RCC Security Officer must be concerned with the well-being of the RCC itself, the physical and cyber threats it encounters and the significance of these threats to the entire MASS fleet under the control of the RCC and the ports they serve. The RCC staff will also perform all planning and scheduling of routine shipboard maintenance with the performance of routine and exception maintenance accomplished under the supervision of senior deck officers. A brief look at these existing and potentially new MASS departments and their functions will highlight several of these issues.

*Engineering* – With the relatively high levels of automation presently existing in shipboard propulsion, power and other engineering systems, the transition to MASS operations in the Engineering Department is likely to lead the charge towards full automation. Looking specifically at the duties and responsibilities of this department, there are several areas where changes may be anticipated.

The Engineering Department on the ship holds the responsibility to ensure that the engine room machinery is in proper working condition to ensure a smooth voyage. Under International Convention on Standards of

Training, Certification and Watchkeeping (STCW) Code requirements it is necessary to ensure the ship's machinery and equipment are working in an efficient manner to support safe navigation of the ship [STCW 95, A-III/2]. This includes the responsibility for frequent inspections of all the equipment dealing with ship, personal safety, fire prevention and pollution prevention at regular intervals of time. Unmanned engine rooms are already the norm on most newly built vessels, but engineers are on board to perform inspections as needed. However, MASS by definition can be operated without seafarers on board, thereby preventing direct, in-person inspection by qualified personnel. It will therefore be necessary to ensure new methods and standards are created to carry out remote inspections and to validate their findings. This requires the use of sensors upon demand and in real time capable of detecting the presence of water, oil and other chemicals where their existence is inappropriate as well as visual, thermal and other indications of fire, flooding and other anomalies that can occur with machinery and equipment. The use of robotic devices will also be necessary to properly position sensors and to manipulate equipment, valves, cover plates, hatches and other devices in their present form and as specifically modified to support robotic manipulation to accomplish inspections.

The use of battery, hydrogen fuel cells, LNG, wind and other new fuels and new propulsion technologies to decrease reliance upon internal combustion engines can dramatically reduce or eliminate the need for maintenance of fuel and lubrication oil purifying equipment and the collection of waste oil. The introduction into MASS of highly reliable and multiple redundant systems along with condition-based maintenance and improved failure prognostics can significantly reduce the need for or eliminate on-vessel maintenance while underway and associated on board spare parts.

The understanding of the MASS engineer of how to respond to emergencies can be supplemented with artificial-intelligence-based assistants that are knowledgeable of the locations and operation of shipboard emergency equipment and other important emergency machinery along with procedures for coordinating their use. Emergency response times can be reduced by automation through simulation and actual performance of emergency drills on the vessel itself, leading to more frequent (and even continuous) practice and comprehensive results, gained efficiencies and improved effectiveness. Throughout this process, in an emergency situation, the Chief Engineer can maintain proper communication regarding the situation of emergency and be co-operative with the Captain so that both Deck and Engine departments function towards bringing the emergency situation under control in the fastest possible time.

*Deck (Navigation)* – MASS navigation encompasses all aspects of a voyage including voyage planning, execution and follow-up after arrival at the destination by the Deck Department. Many automated tools presently assist in both voyage planning and post-voyage debriefing and follow-up, with development independent of MASS for use by all vessels. It can be presumed that this trend will continue, and their functionality and capability will evolve along with new technology and applications. However, the tasks performed by the navigation watch will not likely experience significant change as command and control shifts from the bridge to the RCC. This assumption is based upon the necessity of an equivalent level of safety and functionality to perform bridge functions from a location other than the ship's physical bridge such as a "virtual bridge" under human surveillance at the RCC to be considered as being "manned" under the STCW Code [MSC 100/INF.3]. This requires a high-speed communications link with near-zero latency in exchanging information and faithful replication of all ship responses in real time – not an insignificant requirement that may be susceptible to limited degrees of success. Likewise, the act of maneuvering a ship is the duty of the Master, while under IMO Degree Four Autonomy this will be accomplished by the on board reasoning system. One of the more significant differences between navigation from the ship's bridge and an RCC is that on a ship it is physically impossible for off-watch personnel to leave the ship and still be "on call" and available if necessary, while this is not the case at an RCC.

### 5.2.1.2 RCC Handling of Multiple Ship Operations

Whereas the previous paragraph described the jobs and functions associated with the operation of one MASS entity, here we take a look at the next higher level to see how an RCC may manage simultaneous voyages of multiple MASS under their control with many operating at different degrees of autonomy.

One of the attractions of MASS is the opportunity to gain efficiencies by trimming costs, including the cost for crews. Since most of the shipboard tasks during a voyage have been automated, a smaller crew can reside at the RCC to oversee the voyage. Rather than stand by idle for hours watching displays depicting routine operation on a vessel under autonomous control where nothing out of the ordinary is happening, a widespread assumption has been made that one watchstander may oversee the operation of several vessels at a time. One concept envisions several Operators, each handling up to six vessels, supervised by an Officer of the Watch (OOW) with a relief Operator readily available along with a Captain, Engineer and another spare Operator available as part of a Situation Team to intervene on an exception basis if necessary [MUNIN 2016a]. The role of the Operator is to monitor the operation of several autonomous vessels and providing control through high-level commands that include updating the voyage plan or the operation envelope of the vessel [MUNIN 2016b]. The Engineer is responsible for vessel maintenance and for assisting the Operator to provide technical assistance. A Situation Room Team consisting of the OOW, Captain and Engineer could take over direct remote control of a vessel using a remote replica of the bridge. Details are provided describing an RCC as supporting one hundred vessels with a total of three situation rooms [Rødseth 2014].

In April 2016, Rolls-Royce released its vision of a future RCC where a small crew of between 7 and 14 people would remotely monitor and control a worldwide fleet of vessels [Roll-Royce 2016]. This concept, developed in collaboration with the VTT Technical Research Centre of Finland and the University of Tampere Unit for Computer Human Interaction (TAUCHI), depicts cooperation between operators, local controllers and system specialists in a three-tiered RCC dedicated to worldwide operations, local control and situational analysis and problem solving. In this portrayal there does not appear any reference to traditional maritime chain of command. Rather, tasks are performed based upon function.

The MUNIN and Rolls-Royce approaches provide two distinct views of how MASS control and operation may be accomplished from a shore-based remote control center. However, neither approach appears to consider the relevance and significance of communications and security at the MASS and RCC in the composition of the incident or situation team. The roles of maintenance, ship's business and other staff and support roles must also be considered.

## 5.3 REMOTE CONTROL CENTER FACILITIES

Remote control center facilities must implement all of the functions necessary to execute a safe voyage. This applies not only to the operation of MASS from point A to point B, but also for voyage planning and follow-up. The approach described for the design of RCC facilities considers the concepts put forth in MUNIN project results, by Rolls-Royce and others and expands upon these principles to fulfill functional requirements unique to MASS. This is accomplished using a close facsimile of the modern shipboard command structure modified to take best advantage of ship system automation while respecting the knowledge and experience of seafarers.

For the sake of discussion, RCC functionality will be considered at three levels: Management, Operational and Support along the lines illustrated in Figure 5.1. The Management Level represents the Captain, First Mate/Executive Officer and Chief Engineer and integrates a Corporate Representative to make use of MASS Big Data for shipping company business and to interact directly with the Captain when it makes sense to do so. The Operational Level represents traditional seafarer occupations, as they are adapted to MASS for vessel command and departments. Also at this level are new departments created as a direct result of the unique requirements of MASS. The Support Level represents traditional seafarer jobs that have migrated to the shore as well as new staff positions within the RCC to support MASS-specific requirements.

## 5.3.1 Management Level

The Management Level represents the corporate and command structure for shipping where business requirements intersect with operational needs. The Captain is the interface with corporate authorities in terms of conducting ship's business. With digitalization of the shipping industry efficiencies gained in the processing and elimination of paperwork from the home office or charterer, crew lists and payrolls, accounting and other traditional concerns of Captains, workload in these areas has been greatly reduced or eliminated. Without such administrative distractions the Captain is now free to concentrate on what he or she does best – overseeing the sailing and operation of the vessel with the cooperation of the First Mate or Executive Officer, Chief Engineer and the crew.

In times past the corporate interests of the shipping company pertaining to a specific ship were vested in the Captain, with the activities of multiple Captains and ships monitored from a corporate headquarters or operations center. Since the timeliest information for all ships under the control of the RCC regarding performance, schedule and condition will now be available locally, it is anticipated that corporate representation will be embedded within the RCC itself whereas none was previously available or needed onboard the ship. This corresponds to the MUNIN "All Ships Overview" projection and related information, and the Rolls-Royce Operator Experience (OpEx) "Global Wall of Worldwide Shipping Traffic" [MUNIN 2016a, Slide 32; YouTube 2016].

### 5.3.1.1 Corporate Representative

One or more Corporate Representatives will have access to the broadest level of data coverage and summary information relevant to the shipping enterprise from the vessel itself as well as information derived from the Management, Operational and Support Levels. This is likely to include individual vessel names, positions, ports of origin and destination and time information, schedule, vessel speed, present status and maintenance conditions. It would also include vessel performance data such as fuel consumption, energy reserves, correlation to external factors such as weather and currents, along with trend and other information that may be useful for optimizing future operations. Information could also be presented pertaining to the composition of the RCC staff, their assignments, tasks and workloads, and details of incidents and their resolutions.

The Corporate Representative work station would consist of one or more display screens integrated with a smart chair of standard design from which access to all relevant data products and analysis tools would be accomplished as shown in Figure 5.2. This smart chair would be interoperable with the same chair design across all levels of the RCC including the Captain, First Mate, Chief Engineer, Helmsman, Second and Third

## 98 Unmanned and Autonomous Ships

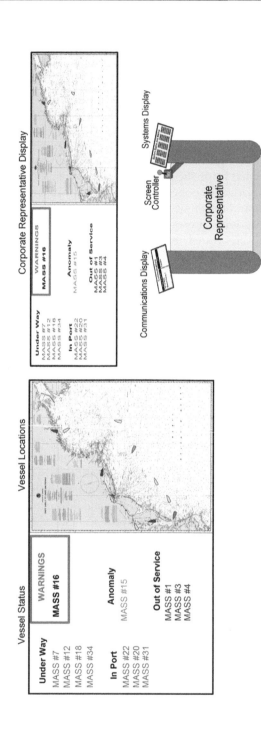

Figure 5.2 Corporate representative chair and visual displays.

Mates and Engineers, technical experts and other positions and certain support staff. However, each work station position will possess its own functionality as appropriate to job tasks and responsibilities. For example, the Captain, First Mate and Helmsman would all have the capability to access and analyze individual vessel datasets, sensor information and even to take over control to navigate the vessel using a joystick or other suitable human-machine interface. Corporate Representatives and Engineers would use this same joystick (or interface device) to access and analyze datasets appropriate to their jobs and responsibilities, but would not be able to navigate the vessel. The arms of each chair would be equipped with Systems and Communications display consoles that provide current status of systems or operations tailored to each job and to provide instant communications with other crew members and staff with whom they must confer and interact on an ongoing basis. This capability would even extend to VHF radio communications with other vessels by the bridge staff as may be needed to facilitate navigation. Such an approach provides the ultimate in RCC system redundancy to reconfigure assets in response to individual equipment failures, to meet surge requirements in the event of bad weather, and to handle incidents.

One or two work stations would be allocated at a RCC for Corporate Representatives as may be needed to perform their duties and to access RCC staff.

### 5.3.1.2 Captain and First Mate

While the Corporate Representative(s) have direct access to status and summary information regarding all ships under his or her cognizance, which may number up to 100 or more, the scope of the Captain and First Mate purview is limited in number to the 36 or so vessels for which he or she is directly responsible as the supervisor for one to six Helmsmen operating up to six vessels each. Summary information made available to the Corporate Representative(s) should also be available to the Captain for each ship under his or her command, only to a much greater level of detail. In addition, the Captain and First Mate should have direct access to the exact information that is available to the Helmsman and that can be accessed as needed to ensure adequate supervision and, if necessary, take over the helm.

Figure 5.3 depicts the navigation displays of the Captain/First Mate (5.3a on left) and the Helmsman (5.3b on right) illustrating the commonality between the positions including Radar, ECDIS, current weather Radar displayed on the auxiliary screen and the navigation Sonar display under the ECDIS display. However, subtle differences exist in the displays due to the different scopes of responsibilities at each position. At the Management Level the Captain/First Mate is responsible for one to six Helmsman along with Second and Third Mates at the Operational Level. On the left side of the Captain's display (Figure 5.3a) is shown exactly what each Helmsman

100  Unmanned and Autonomous Ships

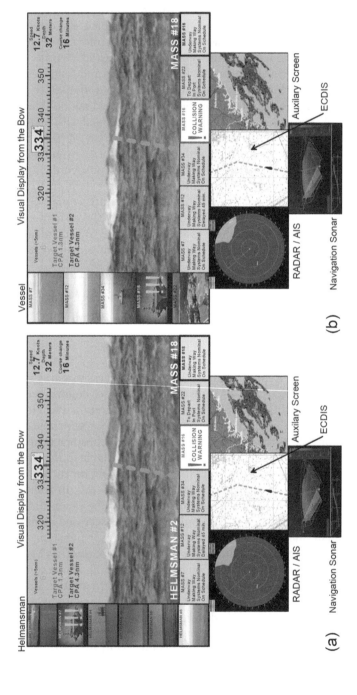

*Figure 5.3* Navigation visual displays. a. Captain/First Mate navigation display. b. Helmsman navigation display.

is seeing and which vessels are being worked. This display will change in real time to correspond with Helmsman activities in supervising each vessel and includes the ability to create an exact replica of the Helmsman's main visual display.

In this example an exception is shown for Helmsman #2 (second from top on left) on the Captain's screen (Figure 5.3a), where the ominous vision of an aircraft carrier approaching MASS #16 from dead ahead is highlighted with exclamation marks, illustrating an anomaly that Helmsman #2 should be aware and energetically seeking to resolve. However, in Figure 5.3b the display of Helmsman #2 shows he or she is currently observing another vessel (MASS #18). The Captain/First Mate would then have the option to intervene directly or to issue a warning to Helmsman #2 to resolve the problem.

The Systems Display on the right arm of the Captain/First Mate chair would be configured to display overall vessel system status while on the Helmsman chair it would only include information appropriate to vessel navigation. The Communications Display for each position is configured appropriate to its level of responsibilities to ensure efficient communications between the crew members.

In addition to the navigation responsibilities of the First Mate the work station for this position would also include capabilities for the display of safety information including the status of all related sensors, exception reports and details regarding incidents related to the safety of the ship and cargo and its safe loading, unloading and stowage on the vessel.

### 5.3.1.3 Chief Engineer

As the highest-level position supervising all aspects of engine and related machinery in the Engineering Department, the Chief Engineer plans the manpower and supervises the Second Engineer and the performance of the tasks and duties of his staff. Within the RCC the Chief Engineer would have direct access to sensor and telemetry data for all essential systems as well as analytical tools to support decision making with planning and maintenance issues. Figure 5.4 depicts a display of the Engineering work station that provides Management Level information regarding all vessels under his or her authority including the condition of engineering systems, anomalies encountered, status of repairs underway and repair history, maintenance and inspection schedules and actions performed, system prognostics and diagnostic data, and other tools to assist in the performance of his or her duties and the supervision of subordinates.

Similar work stations and displays would also be available for Second and Third Engineers configured to assist them in performing specific tasks assigned to them, including analyzing operational system data and deploying robotic devices to perform inspections and maintenance.

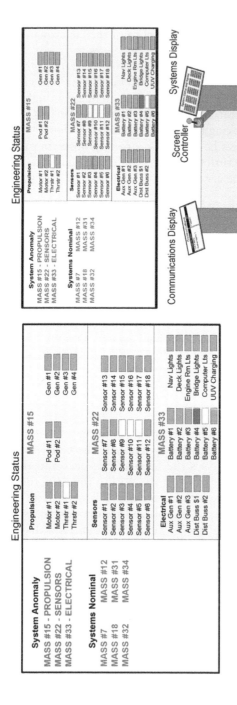

*Figure 5.4* Engineer chair and visual displays.

## 5.3.2 Operational Level

Remote Control Center functions performed at the Operational Level involve Second and Third Mates who traditionally stand navigation watch including acting as Helmsman, along with Second and Third Engineers who stand watch over the engine room and perform inspections to ensure smooth operation during a voyage and to carry out periodic maintenance when in port and at sea using robotic devices.

### 5.3.2.1 Second and Third Mates

The role of the Second Mate at an RCC is to focus on route planning and being the one most familiar with the navigation and watchkeeping systems on the bridge. This includes visual imagery and electronic sensors such as Radar, Automatic Radar Plotting Aid (ARPA), Electronic Chart Display Information System (ECDIS), Automatic Identification System (AIS), depth sounder, navigation Sonar and other devices dedicated to enhancing situational awareness about the vessel and its immediate vicinity. The Third Mate is responsible for the emergency systems onboard the vessel, in addition to providing backup support to the Second Mate related to the navigation and watchkeeping systems. Both of these positions require full Helmsman qualification and the performance of watchkeeping duties. This would typically be accomplished from the First Mate work station position while the First Mate is standing watch in the Captain's chair. A deck officer that is on duty but not standing watch would serve as backup for Helmsman in the event of need.

### 5.3.2.2 Helmsman

When MASS operations are performed under IMO Degree Three Autonomy the vessel is remotely controlled and does not have any crew on board. This control is performed using the Navigation Display by the Helmsman who is responsible for maintaining a steady course and properly executing rudder orders through communication with the officer on the bridge. Operation in this mode presupposes a high degree of attention being paid to performing this task while having responsibility for only a single vessel. An exception may be made in the event a vessel is on autopilot while transiting open waters with little or no traffic expected to be encountered along the route. When operating under IMO Degree Four, fully autonomous mode the Helmsman follows the actions of the MASS automated reasoning system performing this task to ensure it is being accomplished properly.

The chair and visual displays used by the Helmsman at the RCC to perform this task are illustrated in Figure 5.5. The Main Display would typically provide an unobstructed visual rendition of the path of transit forward the bow with an overlay of appropriate navigation information such as course, speed, time of next course change. This rendition would

104  Unmanned and Autonomous Ships

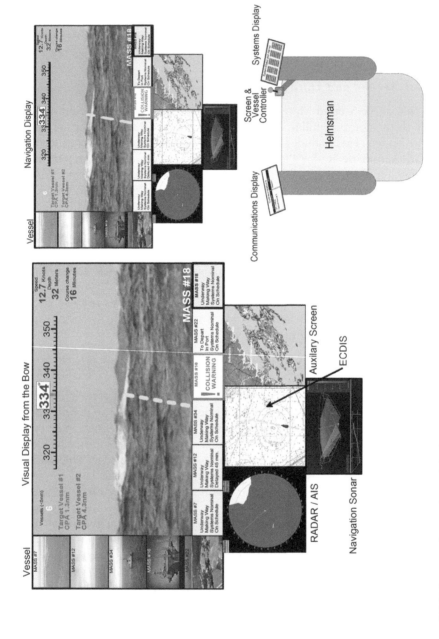

Figure 5.5  Helmsman chair and visual displays.

typically consist of live video streamed imaging from a bow camera capable of being augmented with infrared, Radar, Light Detection and Ranging (Lidar), navigation Sonar and/or other imaging technologies as may be appropriate during night or low visibility conditions. It would also display visual indications of objects detected and recognized within the view such as other vessels, aids to navigation, hazards to navigation such as marine mammals, shipping containers and large objects adrift.

Below the Main Display lies the navigation instrument cluster on which the Helmsman relies to steer a proper course, detect other vessels and buoys and to perform all tasks required of this position. This instrument cluster is greatly simplified as compared with many modern navigation systems and is arranged in an ergonomic fashion with ECDIS at the center, Radar with ARPA on the left of ECDIS, navigation Sonar below ECDIS and an auxiliary screen on the right of ECDIS displaying supplementary information essential to safe navigation that can be changed in real time as needs require throughout the voyage. Radar and AIS overlay on ECDIS should also be supported to make use of their added value through their fusion of imagery, and especially to detect inconsistencies and errors that may be present in Electronic Navigation Chart (ENC) source data [MSC.1 2017]. These instruments are purposely grouped together in a small physical area to allow the Helmsman to always look at meaningful information without having to turn his or her head, and not get sidetracked by nonessential information that wastes time, causes distraction and contributes to poor performance. This is contrasted with many modern bridge configurations without standardized layouts with two or more large display monitors crammed with hundreds of data items, few of which are essential to visually monitor on a continuous basis.

On the left side of the Main Display are the Vessel Screens containing from one to six live images representing the view along the path of transit forward the bow (default case) obtained from other vessels under the supervision of the Helmsman. This image would automatically be changed to one of greater significance should the onboard reasoning system detect a hazardous condition or threat, with appropriate notification given as to the nature of the condition. The Helmsman would then be able to transfer his or her control to the vessel where appropriate corrective action may be taken.

Along the bottom of the Main Display are the Status Screens representing summary information pertaining to navigation capability for each of the vessels being supervised. In addition to the information content the background of these screens can be color coded to reflect nominal conditions, warnings and immediate action requirements (e.g., green, yellow and red) along with an indication as to which vessel's information currently occupies the Main Display (e.g., a blue background).

The Communications Display on the left arm of the Helmsman chair would be configured for instant communication with the bridge officer using earphone and microphone, and to monitor VHF radio traffic from other vessels and to initiate communication of intentions between vessels during close maneuvering and crossing situations. The Systems Display on the chair right arm would show the status of navigation, imaging and other instruments essential for navigating the current vessel depicted on the Main Display.

The Screen and Vessel Controller is comprised of a joystick or other human-machine interface that enables the Helmsman to change control to any vessel under his or her purview by selecting one of the ships displayed on the Vessel Screens. Upon selection, the Main Display is populated with imagery and sensor data for the selected ship and the previous ship is returned to the Vessel Screen. Simultaneous with this action the joystick, in combination with other controls existing within the displays, can now be used by the Helmsman to steer, control the speed and make other adjustments to the vessel propulsion and control surfaces needed to perform safe navigation.

### *5.3.2.3 Second and Third Engineers*

The Second Engineer is responsible for supervising engine room operations and daily maintenance of engine room and associated equipment for all of the vessels for which he or she is responsible. This will involve the simultaneous supervision of multiple vessels which may or may not be the same (up to) six vessels under Helmsman control. Future experience in handling multiple MASS vessels will dictate the manageable distribution of vessels between engineers.

As transitioned to MASS this position will require the capability to remotely inspect the engine room and propeller configuration using cameras, remote sensors, robotic and other devices. Such capability will be contained within the common Engineering work station designed for use by all engineering positions. The actual performance of inspection, maintenance and repair activities will be accomplished by the Third Engineer who will also stand engineering watches. The implementation of the engineering watch involves intricate knowledge of the equipment and systems to make sense of what the instruments are indicating, combined with analytical tools and innate touch, sight, smell, hearing and other senses to prevent or detect, identify and correct system anomalies that may interfere with vessel operation. These senses must be replicated in the engine room and integrated within engine room machinery through the placement of sensors and the use of robotic devices to perform inspections and maintenance.

### *5.3.2.4 Security Officer*

The Security Officer is customarily appointed by the shipping company in consultation with the Captain to help ensure the security of the ship and her crew. In the case of MASS without crew this task must be performed remotely, yet the threats remain the same. These include attempted boarding, armed robbery and piracy, hijacking, stowaways, sabotage and terrorism, as well as an assortment of potential cybercrimes. Further complicating the matter is that in addition to physical and cyber attacks to the ship itself, MASS is also susceptible to a completely new set of vulnerabilities. These include assuming control by making the vessel think it is the RCC issuing commands when it is in reality an entity pretending to be the RCC that is blocking or overriding legitimate commands and signals. This can occur by hacking the RCC itself and taking over its systems to control one or more MASS directly, or by stationing a vessel in close proximity to the MASS vessel to interfere with and overcome its normal command and control channels and substituting another. A more subtle form of hacking MASS will be through unsecured or improperly secured Internet of Things (IoT) devices on the vessel itself including cameras, computers, device controllers, displays, etc., that can provide a pathway to obtain sensor data and direct access to computers to allow for the introduction of malicious software for future activation.

While the Vessel Security Officer has detailed knowledge of how to thwart and overcome physical intrusion and breach of vessel defenses by outside forces, the nature of cybercrimes and hacking requires a completely new and comprehensive skill set – and especially in the case of MASS. For this reason there needs to be two separate functions in terms of the Vessel Security Officer and the RCC Security Officer, both of whom should reside within the RCC to effect rapid reaction to threats. These functions must consider the ability to detect and respond to both physical and cyber threats that occur independently. A greater threat is coordinated cyber and physical attacks that blind the vessels defenses to not recognize real threats and essentially open the door for physical assault. This can be accomplished by using "deepfake" artificial intelligence techniques to substitute computer-generated imagery and data for the real imagery and data feeds from MASS sensors.

The Vessel Security Officer would be responsible for maintaining security for a select number of ships and notifying the RCC Security Officer of threats and potential threats to MASS in port and while underway. The position would require intricate knowledge of the various devices and imaging platforms installed in MASS including perimeter Sonar, video and infrared cameras, microphones and other sensors capable of detecting threats and even recognizing specific threat situations. It would also require an operating knowledge of MASS that would facilitate the detection of denial of service and spoofing attacks that may render navigation systems ineffective or even dangerous and susceptible to attack and highjacking. Ideally MASS would also be constructed so as to be capable of securing the vessel from outside forces short of a direct attack that could breach the hull. Even if the propellers could be disabled effectively rendering the vessel dead in the water, prior knowledge of an imminent attack could buy time sufficient to render assistance.

The RCC Security Officer would be tasked with maintaining physical security and cybersecurity of the Remote Control Center itself, including the outside perimeter as well as all internal access points to MASS crew and support staff. This position must work closely with Vessel Security Officers to keep them informed of potential threats they may be faced with from remote operations including data breaches, insertion of malware and software trojans, and attempts to render control of navigation and/or engineering work stations and other resident computer systems.

### 5.3.2.5 Communications Officer

The data pipeline between MASS and the RCC provides the lifeblood through which all command, control and knowledge of remote and autonomous operations is passed. At one end is the vessel itself that generates and receives gigabytes of command instructions, telemetry data and imagery on an hourly basis, which could be considered an IoT device consisting of distributed subsystems of other IoT devices. At the other end is the RCC that uses the data products generated from MASS data and imagery to provide situational awareness to the Deck and Engineering Departments in IMO Degree Three remote operation mode and, in addition, supervise the onboard reasoning system while operating in during IMO Degree Four full automation mode. In between are various terrestrial and satellite communications channels through which these data and imagery pass, cloud computing and storage used to process these data and imagery and to implement digital twin simulation of MASS reasoning capability to ensure consistency and predictability of automated reasoning system outcomes.

The Communications Officer is a new MASS crew position responsible for knowing the details regarding all parts of the data and imagery pipeline, the equipment that implements its functionality, the limitations that may affect bandwidth and restrict operations, and the vulnerabilities of these systems to interruption and unauthorized penetration. Specific knowledge of 5G satellite and cellular systems is required along with terrestrial Internet feeds and their use to provide communications channel redundancy and backup capability in the event one or more of these services are unavailable. Reporting is direct to the First Mate or Captain, in coordination with the RCC and Ship Security Officers as required to maintain awareness of cybersecurity threats and disruptions to satellite, cellular and terrestrial communication channels.

### 5.3.2.6 Automation Officer

Machine learning and artificial intelligence deep learning reasoning systems form the core of automation technologies for the Deck and Engineering Departments. There exists a need for someone knowledgeable about the onboard reasoning systems to report to the Captain and be able to explain how and why these systems make their decisions and perform as they do during actual operation as well as during simulations. This requires intimate knowledge of the machine learning methods, neural network architectures, training sets used for development, testing, verification and validation of these systems. In addition, the knowledge gained of how the automated reasoning systems perform in the real world can be applied directly to retraining and refining the training processes used by these systems to improve their performance, reliability and resilience.

The Automation Officer would also be responsible for supervising digital twin simulations being conducted in real time using cloud computing or at the RCC itself using the same exact sensor and other data as the onboard automated reasoning system to ensure similar and functionally equivalent results are attained by both systems. In the event different results occur; the Automation Officer and support staff would identify their cause(s), establish the rationales for their occurrence, perform actions required to resolve the problem and perform follow-up development, testing and installation of automated reasoning software.

#### 5.3.2.7 Unmanned Air Vehicle (UAV)/Unmanned Underwater Vehicle (UUV) Operator

Another new position to result from MASS but which can apply to all surface vessels is that of a UAV/UUV Operator. In addition to any specific national requirements for drone operators, additional requirements should exist regarding knowledge and experience in maritime applications and use of these technologies in qualifying for privileges for maritime operations. These requirements would not only apply to technical qualifications for operating drones, but for rules and regulations as may be defined for their safe operation in a maritime environment.

### 5.3.3 Support Level

Support to MASS crews in the form of readily available technical experts and administrative assistance at the RCC is a luxury that is generally not available on board conventional ships. Integration of appropriate support personnel into the daily operations can increase crew efficiency and improve overall MASS system performance.

#### 5.3.3.1 Technical Experts

The scope of system automation, sensors and sheer complexity of MASS systems requires establishing local or virtual access to a range of technical experts sufficient to resolve technical problems in a timely manner and to confer with on the proper use and maintenance of and upgrades to equipment and software. They should also be knowledgeable of known problems and deficiencies and be able to recommend and devise workarounds to overcome these deficiencies while permanent solutions are sought. These experts may include manufacturers' representatives as well as shipyards and others knowledgeable of specifics of equipment design and performance, vessel construction and the installation of equipment on the vessel. Similar requirements exist for supporting the systems and equipment at the RCC itself.

#### 5.3.3.2 Other Support Staff

Administrative support should be made available to Corporate Representatives and Management Level crew members to assist with, oversee and coordinate administrative procedures and follow up on administrative tasks initiated within the RCC. This can include entering data, preparing and distributing reports and managing company records.

## 5.4 REMOTE CONTROL CENTER ORGANIZATION

The RCC must support all MASS functions, the crews who operate them and the shipping companies themselves in their ability to operate the enterprise in an efficient manner.

### 5.4.1 Functional Organization

One possible implementation of the functional organization of the RCC to ensure close cooperation and interaction across all responsibility levels is shown in Figure 5.6, which describes one cohesive unit of the RCC organization. The entire RCC organization may consist of one or more of these Functional Units to meet the needs of an expanding workload. This approach will implement all authority levels and provide great flexibility in operations to meet changing needs and priorities on a daily basis and in the long term.

The primary focus of the RCC must be the vessels themselves and the ships' crews that operate them. Closely allied are the corporate representatives who benefit from close coordination and cooperation with the Captains and crews and access to the vast amounts of data and information necessary to achieve efficient and profitable operations. The proposed approach to organizing a RCC is to designate clear lines of authority and responsibility by functional makeup, with deference to upholding maritime traditions and practices. The continued use of traditional position names such as Captain, First Mate, Engineer, Able Bodied Seaman, etc., along with the responsibilities of each was described earlier in this chapter. Their implementation as a unit that is part of the functional makeup of the entire RCC is considered in this paragraph.

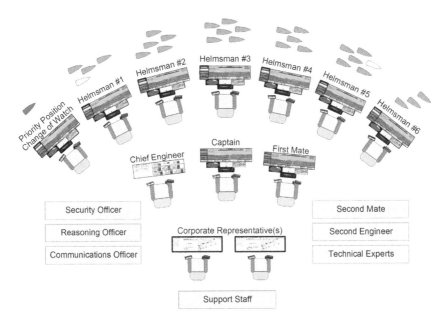

*Figure 5.6* Organization of one functional unit of a Remote Control Center.

The makeup of the RCC Functional Unit represents 16 or more individuals handling a total of 6 to 36 vessels at a time. The presumption of one Helmsman being capable of handling between one and six vessels is based in part on the results of previous studies from MUNIN, Rolls-Royce and other efforts for RCC design and capacity. During normal, routine operations, a Helmsman could be assigned up to six vessels operating under IMO Degree Four fully autonomous mode. However, IMO Degree Three remote operation from the RCC would warrant more limited scope to one, two or possibly three vessels depending on the nature of the voyage and workload required by the circumstances. The handling of MASS in extremis would warrant the undivided attention of one Helmsman. The RCC Functional Unit would consist of a maximum of six Helmsman and seven chairs, with the seventh chair reserved for change of watch allowing overlap between Helmsman to confer between watches, The seventh chair could also be used as a backup chair for a Helmsman or Deck Department officer to rapidly intervene should the need arise.

The Captain, First Mate and Helmsman each utilize a standardized, identical chair. However, the Deck Department display screens for the Captain and First Mate are configured to supervise up to seven Helmsman chairs. Second and Third Mates also use the same standard chair configured for the specific aspects of their jobs and responsibilities. The Chief Engineer, Second and Third Engineers all use the standard chair and basic Engineering Department display, each of which is configured to correspond with their specific jobs and responsibilities. The Security, Communications, Automation Officers and Corporate Representative(s) all use the standard chair. However, their display screens are tailored to the specific needs corresponding to their unique functions. Technical experts and support staff have unique requirements of their own and may or may not require the use of a standard chair.

### 5.4.2 Allocation of Physical Space

The allocation of physical space within the RCC facility should consider locating the Corporate Representative function independently at a level that makes possible rapid communication and direct interaction with the Captains yet does not promote distraction and interference with their tasks. Concerns regarding any of the vessels under control of the RCC can be brought directly to the attention of the responsible functional unit Captain who may or may not already be aware of their concerns depending on relevance to their primary mission of safety of navigation and ship performance in vessel operations.

The First Mate and Chief Engineer work stations should be in close proximity to the Captain on the "virtual bridge" to promote rapid, in-person communication between departments and to confer with the Captain as necessary. Second and Third Mate and Engineer positions can be located remotely from the bridge but keep in constant voice contact through earpiece and microphone and message contact through their Communication Displays.

### 5.4.3 Computational Facilities

The massive amounts of sensor data and imagery generated by MASS must be rapidly processed and analyzed, with results quickly disseminated to the crews, corporate and support staff located at the RCC and to corporate headquarters. It is likely that the most efficient and cost-effective solution to accomplish this task is through the use of cloud computing. This approach takes advantage of the greatest possible data transfer rates and parallel computing capabilities rather than investing in high bandwidth communications infrastructure and massive computing facilities to be located within the physical RCC facility itself. This would also extend to cloud-based, digital twin reasoning simulations to replicate on board MASS reasoning functions using the exact same sensor data being used on the vessel itself. However, there is a danger that cloud computing and technical support may not be available using the same exact hardware and software configurations as exists onboard MASS. Further danger exists in determining the combined latencies of the communications channels for transferring extremely large amounts of sensor data and imagery as well as the time required for cloud computing to formulate and report results to the RCC.

In such cases, provisions must be made for accomplishing digital twin simulation and the reporting of processing results within the RCC itself.

**REFERENCES**

Aamaas, Paal. 2018. USN Signs Agreement on Development of Shore Control Center – For Autonomous Vessels, 17 October 2018. https://autostrip.no/usn-signs-agreement-on-development-of-shore-control-center-for-autonomous-vessels/.

Baldauf, Michael, Raza Ali Mehdi, Momoko Kitada and Dalaklis Dimitrios. 2017. Shore Control Centres for Marine Autonomous Systems: Exploring Equipment Options Marine Autonomy & Technology Showcase [MATS] 2017 Presentation, November 2017. DOI: 10.13140/RG.2.2.29964.41605.

MSC.1 2017. ECDIS – Guidance for Good Practice. MSC.1/Circ/1503/Rev.1. Maritime Safety Committee. International Maritime Organization. London, 16 June 2017, p. 15.

MSC 99/INF.3. Final Report: Analysis of Regulatory Barriers to the use of Autonomous Ships. Denmark, p. 6.

MSC 100/INF.3. Initial Review of IMO Instruments under the Purview of MSC. Note by the Secretariat. Regulatory Scoping Exercise for the use of Maritime Autonomous Surface Ships (MASS) Scoping Exercise. STCW. Annex, p. 39.

MUNIN 2016a. Navigational Shore Support: A New Perspective. The Shore Control Centre. Unmanned Ships Maritime Unmanned Navigation through Intelligence in Networks (MUNIN). MUNIN-Final-Event-B-4-CTH-MUNINs-Shore-Control-Centre. 2016. www.unmanned-ship.org.

MUNIN 2016b. MUNIN Results. 2016. http://www.unmanned-ship.org/munin/about/munin-results-2/.

North 2015. *VDR & SVDR. Loss Prevention Briefing*. The North of England P&I Association, Newcastle upon Tyne, UK. www.nepia.com/media/869615/VDR-LP-Briefing.pdf.

Ottesen, Are E. 2015. Situation Awareness in Remote Operation of Autonomous Ships: Shore Control Center Guidelines. www.ntnu.no/documents/10401/1264435841/Artikkel+Are+E+Ottesen.pdf/abb533ae-e73a-489e-80ec-f0e198e72c0a.

Rødseth, Ørnulf Jan. 2014. Control Centre for Unmanned Ships. Presented to "Regionalt nettverk for kontrollrom, sikkerhet og fjernstyring". Trondheim, 6 March 2014. https://www.sintef.no/globalassets/project/hfc/documents/munin-scc.pdf.

Roll-Royce 2016. Rolls-Royce Reveals Future Shore Control Centre. Press Release, 22 March 2016. https://www.rolls-royce.com/media/press-releases/2016/pr-2016-03-22-rr-reveals-future-shore-control-centre.aspx.

SOLAS V. Safety of Life At Sea Convention, Chapter V, Regulation 20. Voyage Data Recorders.

STCW 95, A-III/2 Standards of Training, Certification and Watchkeeping for Seafarers. International Maritime Organization (IMO). Parts 3–2. 53. London. 1995.

YouTube 2016. Rolls-Royce – Future Shore Control Centre. 23 March 2016. https://www.youtube.com/watch?v= vg0A9Ve7SxE.

# Chapter 6
# Navigation

Methods of navigation for future unmanned and autonomous ships can be described in many ways. One characterization of how a remotely controlled vessel may operate is stated as follows:

> "The captain with a giant screen which overlays the environment around his vessel with an augmented reality view can navigate confidently using the computer-enhanced vision of the world, with artificial intelligence spotting and labeling every other water user, the shore, and navigation markers."
>
> [Stewart 2018]

This concept can be expanded to describe a fully autonomous ship merely by replacing the captain with an automaton. The boldness, self-assurance, confidence and trust exhibited by such a statement in the capability and correctness of the systems on which the captain must rely is at present entirely premature as the safe and reliable performance of such systems has yet to be proven. Furthermore, comprehensive below the waterline situational awareness is not even mentioned in this example, illustrating a key weakness in the regulatory framework for modern shipping despite the availability of commercial off-the-shelf instruments able to provide such capability at relatively low cost.

This chapter describes key concepts related to vessel navigation using physical and electronic aids to determine current and future position and other vessel traffic to avoid collisions, allision with objects and groundings and to handle other contingencies that may be encountered during transit. Also discussed are the various sensors and sensor systems needed to survey the physical environment and generate imagery and signals to recognize and identify navigation aids, other vessel traffic, land masses and hazards to navigation as well as the limitations of such systems. Reasoning processes that analyze sensor information to determine object and event significance to achieve safe transit from a point of origin, along a predetermined route, to the destination are also described. Consideration is given to deviations, interruptions and changes to a navigation plan necessitated by other vessel traffic, adverse weather conditions and equipment malfunctions as well as hostile or other actions committed against an unmanned and autonomous vessel itself and against the sensor systems on which vessel navigation relies.

## 6.1 AIDS TO NAVIGATION

The use of Aids to Navigation (ATON) to guide travelers toward their destinations is commonplace across all modes of transportation. Examples include simple arrows carved into tree trunks along a foot path, signs imprinted with the names of towns, lines along the centers of roads to designate lanes of travel, numbers painted at the ends of runways to indicate their direction on the magnetic compass, and stars in the night sky to guide the navigation of caravans on land and ships at sea. In modern times these concepts have been extended through the use of mechanical and electronic systems to provide greater precision in guiding aircraft along flyways and landing in adverse weather conditions, traffic lights to control vehicle passage at intersections, and light ranges to guide ships through a channel.

Marine ATON are devices or systems external to vessels designed and operated to enhance the safe and efficient navigation of vessels and/or vessel traffic [IALA 2014]. They generally consist of a broad range of things and entities such as natural landmarks that include mountains, river mouths, islands, bluffs and prominent landscape features and characteristics; significant manmade objects in the landscape including towers, buildings and roads; and systems of buoys, beacons, ranges, lighthouses and other aids designed specifically for marine navigation. Vision sensors onboard an autonomous vessel would ideally be capable of imaging ATON with sufficient resolution to detect their characteristics, make a positive identification and determine their position through the use of Global Navigation Satellite System (GNSS), Electronic Chart Display Information System (ECDIS) and other methods. Visual sensors may be supplemented with Radar and navigation Sonar to confirm ATON positioning on ECDIS with real-time observations. ATON transmitted using Automated Identification System (AIS-ATON) that may or may not be co-located with physical ATON, which are viewable on AIS receivers onboard the vessel, provide another means for determining position. Virtual ATON (VATON) that require no physical infrastructure can also aid in determining position as a backup to GNSS, ECDIS and the vessel echosounder used to provide navigation through contour tracking along the seabed [Wright and Baldauf 2016d]. VATON may be placed worldwide where physical and AIS-ATON are not possible due to harsh environmental conditions and where remoteness of location restricts or prevents adequate installation and maintenance.

Current trends in unmanned and autonomous shipping are geared toward using to the best advantage possible existing ATON for navigational guidance. However, the unique requirements associated with machines performing navigation tasks absent direct human involvement are likely to spawn new needs in terms of how to designate significant waypoints geo-referenced to the physical environment along seaways and within ports as well as guiding vessels with great precision while approaching wharfs and quays. The need for and the advantages of enhanced physical and signal characteristics integrated within the designs of physical ATON and for new types of Virtual ATON was recently demonstrated under operational conditions in active waterways [Wright 2019a, 2019b]. Further innovation in the very near term is also likely in the development and application of sensors to be used to perform these tasks.

## 6.2 COLLISION AVOIDANCE

The International Regulations for Preventing Collisions at Sea (COLREG) require

> every vessel shall at all times maintain a proper look-out by sight and hearing as well as by all available means appropriate in the prevailing circumstances and conditions so as to make a full appraisal of the situation and the risk of collision.
>
> [COLREG Rule 5]

The proper use of Radar equipment is also required to obtain early warning of risk of collision, and to use Radar plotting or equivalent systematic observation of detected objects [COLREG Rule 7b,c]. These same regulations warn mariners that assumptions shall not be made on the basis of scanty information which, in the case of Maritime Autonomous Surface Ships (MASS), makes it obligatory for the International Maritime Organization (IMO) and member states to ensure adequate sensor capabilities exist for remote operators and onboard reasoning capabilities to have all information needed to ensure they are fully informed as to their circumstances and capable of making all decisions properly.

These regulations were written for vessels staffed by seafarers where reliance on the human senses and human interpretation of environmental conditions, navigation charts and instruments based upon their own knowledge and experience is essential to the execution of a safe voyage. The regulatory framework is presently limited to human vision and hearing and to instruments such as the single-beam echosounder, Radar, AIS, ECDIS and GNSS to fulfill these requirements. However, human senses are absent and modern instruments required for carriage on vessels fall far short of ensuring safety of navigation by remotely controlled or autonomous vessels. Under present international regulations MASS research and development is currently limited to within national waters and between adjacent countries. The IMO has conducted a regulatory scoping exercise to examine and amend the regulatory framework to enable the safe, secure and environmental operation of partly or entirely unmanned MASS and their interaction and co-existence with manned ships within the existing IMO instruments [MSC 98/20/2]. This may or may not result in new equipment carriage requirements for MASS. However, deliberate consideration of the potential for including new types of vision and environmental sensors from all perspectives must be made to ensure safety of navigation.

One example of research performed to examine issues associated with collision avoidance for autonomous vessels was conducted on the North Sea on 19–20 March 2019 near Den Helder in the Netherlands under the Joint Industry Project Autonomous Shipping. This was conducted using *SeaZip 3*, a Damen fast crew supplier vessel from SeaZip Offshore Services, that was equipped with collision avoidance technology to determine how the vessel would interact with seagoing traffic [MAREX 2019]. The autonomous system was connected to the onboard autopilot and machinery control system and performed evasive maneuvers safely. The conclusions reached from these experiments included that further development of autonomous systems was needed to cope with complex marine traffic situations in a more efficient way.

## 6.3 ENVIRONMENTAL SENSOR SYSTEMS

A large number of environmental sensor systems dedicated to monitoring the surface ship maritime environment are needed to not only replicate the human senses, but to exceed human capabilities in terms of their effective physical and spectral range, sensitivity and resolution. Such sensor systems, many of which are illustrated in Figure 6.1, are generally available from four perspectives: the air, water's surface, below sea level and from space. Surface and subsea sensor systems by and large provide real-time ship-centric, line of sight data and imagery while space-based systems provide access to data, information and imagery available from a wide variety of worldwide sources external to the vessel.

The scope and range of sensors needed to safely navigate MASS along long stretches of relatively low traffic deep ocean routes do not differ much from navigation in shallow, coastal waters among archipelagos crowded with both working and recreational vessels. Of much greater significance is the need and ability to properly integrate multiple sensor modalities with reasoning about the vast amounts of available data and imagery. This is necessary to create the information needed to make and explain observations, to critically assess their meaning in terms of mission objectives and

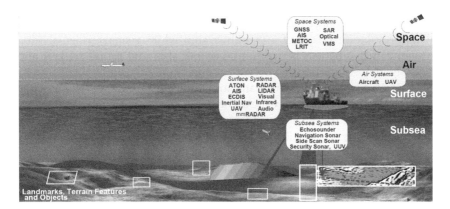

*Figure 6.1* Maritime environment sensor system perspectives.

potential threats, and to plan and execute appropriate responses considering vessel capability and performance. The extent to which MASS can successfully react to sensor observations to minimize risk, to ensure the voyage is completed safely, to recover from dangerous situations and, in the event recovery is not possible, to effectively preserve life, property and the environment will be the ultimate test of MASS technology.

### 6.3.1 Conventional vs. Smart Sensors

The varieties of maritime conventional and smart environmental sensors available from all perspectives are many, and distinctions must be made between these two different types of sensors in terms of fusing and interpreting the data and information they produce in a form suitable for MASS navigation. By far the most common of sensors available to mariners are conventional in nature and merely provide data and imagery through a user interface and/or data bus that must be analyzed on a standalone basis by a computer processor or human. Indeed, today's sophisticated electronic systems are replete with data analytics and even complex reasoning capabilities that can detect schools of fish within the water column and differences between seabed compositions, discern the smallest of vessels and buoys within highly cluttered seascapes, and even view buoy and small vessel identification letters and numbers from long distances. Properly trained users of these instruments can view their outputs and apply this information to their best advantage in planning and executing voyages.

Smart sensors are much like the cameras on smart cell phones that can recognize a human face within an image and adjust the camera's settings to ensure a crisp photo of the person is taken. Smart maritime environmental sensors contain processing capabilities that can analyze the data and imagery obtained from conventional sensors to determine their contents, communicate this information and present it along with any relevant metadata to a human user or intelligent navigation reasoning system where this information may be applied. For example, a video imaging system can provide a high-resolution image of a container ship while a smart video-sensing system can distinguish between a container ship and an oil tanker, and possibly identify the specific container ship being imaged by analyzing its silhouette and even reading its name and home port on the stern. The same example can be given for viewing an image of a navigation buoy and having the mariner identify the buoy, and an onboard sensor processor determines that the buoy is the number 9 green can buoy that is approximately 75 meters from its charted position. Smart sensing of side-scan Sonar images with proper training can recognize and identify specific configurations and sizes of rock formations along the bottom and discern exact vessel location, rather than show an image of rocks and having mariners determine for themselves their location.

The significance of these differences in terms of MASS is where the data, sensor and image analytics will be performed followed by the fusion of this information to form an overall picture of situational awareness. This is also one of the critical challenges of MASS. The preferred configuration contains networks of interoperable, intelligent and distributed sensors providing information directly to an intelligent navigation reasoning system(s) that fuse the information whereby decisions may be made. The alternative network scenario comprised of conventional sensors requires separate processing of disparate data and imagery to create the information to be communicated to the intelligent navigation reasoning system(s). These differences in data and imagery handing and analysis result in a much greater likelihood to introduce error and noise into the data itself, rendering intelligent navigation reasoning systems less reliable and more prone to error and mistakes. The sensors and the fusion of sensor data and imagery to create information for use by intelligent navigation reasoning processes onboard vessels and land-based operators are described herein. The methods and techniques used to analyze and reason with this information and to take all appropriate action are covered in Section 6.4.

### 6.3.2 Shipboard Sensors

Sensor capabilities for use onboard MASS must not merely replicate the sight, hearing and feeling of seafarers, but must exceed their abilities by enabling constant vigilance through 360 degrees around the vessel in four dimensions (x, y, and z coordinates and time) at higher resolution and with greater attentiveness and accuracy than is humanly possible. Many such instruments are illustrated in Table 6.1. This includes the ability to see in the dark in all weather conditions including heavy rain, snow and through fog over the water's surface.

Also needed is the ability to see underwater ahead and around the vessel to detect and respond to nefarious threats and threats not charted, avoid

Table 6.1 Sensor Types and Data Classes among Maritime Surface, Subsea and Space Systems

| | Sensor Type | Data Class | Data Domain | | | | | Data Content | | | |
|---|---|---|---|---|---|---|---|---|---|---|---|
| | | | Pixel | Time[A] | Freq.[A] | Unique ID | Position | Ground Track | Speed | Other |
| **Surface Systems (Shipboard)** | | | | | | | | | | |
| Aids to Navigation (ATON) - Physical | Receiver | Imagery | YES | NO | NO | YES | YES | NO | NO | light/sound |
| Aids to Navigation (AIS) - AIS | Receiver | Data | NO | YES | NO | YES | YES | NO | NO | |
| Aids to Navigation (VATON) - Virtual[B,C] | Data Object | Data | NO | NO | NO | YES | YES | NO | NO | |
| Automated Identification System (AIS) | Transceiver | Data/Imagery | YES | YES | NO | YES | YES | YES | YES | much data |
| Electronic Chart Display Info System (ECDIS) | Data Object | Data | YES | YES | YES | NO | YES | YES | NO | much data |
| Inertial Navigation[C] | Instrument | Data | NO | YES | YES | NO | YES | YES | YES | |
| Laser Imaging (LIDAR)[C] | Instrument | Imagery | YES | YES | NO | NO | YES | YES | YES | |
| Marine Radar (X/S band) with ARPA | Transceiver | Imagery/Signal | YES | YES | YES | YES | YES | YES | YES | |
| millimeter Radar[C] | Transceiver | Imagery/Signal | YES | YES | YES | NO | YES | NO | YES | |
| Visual (video)[C] | Receiver | Imagery | YES | YES | YES | Indirect | YES | Indirect | Indirect | much data |
| Infrared (IR)[C] | Receiver | Imagery | YES | YES | NO | Indirect | YES | Indirect | Indirect | much data |
| Audio (sound) | Receiver | Signal | NO | YES | YES | Indirect | YES | Indirect | Indirect | |
| Unmanned Aerial Vehicle (AUV)[C] | Receiver | Imagery | YES | YES | YES | Indirect | YES | Indirect | Indirect | much data |

(Continued)

Table 6.1 (Continued) Sensor Types and Data Classes among Maritime Surface, Subsea and Space Systems

| | Sensor Type | Data Class | Data Domain | | | | Data Content | | | |
|---|---|---|---|---|---|---|---|---|---|---|
| | | | Pixel | Time[A] | Freq.[A] | Unique ID | Position | Ground Track | Speed | Other |
| **Subsea Systems (Shipboard)** | | | | | | | | | | |
| Echosounder | Transceiver | Imagery/Signal | Some | YES | YES | NO | YES | YES | NO | bottom |
| Navigation Sonar[C] | Transceiver | Imagery/Signal | YES | YES | YES | Indirect | YES | Indirect | Indirect | bathymetry |
| Side Scan Sonar[C] | Transceiver | Imagery/Signal | YES | YES | YES | Indirect | YES | Indirect | Indirect | water column |
| Unmanned Underwater Vehicle (UUV)[C] | Receiver | Imagery | YES | YES | YES | Indirect | YES | Indirect | Indirect | much data |
| **Space Systems (Remote)** | | | | | | | | | | |
| Automatic Identification System (AIS)[C] | Receiver | Data/Imagery | YES | YES | NO | YES | YES | YES | YES | much data |
| Global Navigation Satellite System (GNSS) | Receiver | Data | YES | YES | YES | n/a | YES | YES | YES | much data |
| Meteorological and Ocenographic (METOC) | Receiver | Data/Imagery | YES | NO | NO | n/a | indirect | NO | NO | |
| Optical Imaging[C] (non-METOC) | Receiver | Imagery | YES | NO | NO | n/a | indirect | NO | NO | |
| Synthetic Aperature Radar (SAR)[C] | Receiver | Imagery | YES | NO | NO | n/a | indirect | NO | NO | |

Source: Wright, 2019a.

Notes: [A]Time and Frequency domain data to supplement imagery. [B]Experimental technology not yet in use. [C]Not included in IMO carriage requirements.

groundings and allision and to help avoid whales and other large marine mammals. The capability to hear and measure sound is essential to detect ship whistle and audible signals, ATON sound signals and sounds in the environment such as human voices, wind and waves crashing on rocks. Perception of feeling is needed to detect variations in roll, pitch and yaw attributable to changing sea states, ship behavior from center of gravity movement, and symptoms of damage and failure of ship's equipment. The ability to detect both individual events and series of events over extended periods of time is needed to effectively reason with and apply this information within the context of MASS operations.

IMO equipment carriage requirements related to shipboard sensing augments human sight and hearing with binoculars and hailer listening capabilities. These required sensors perform very well in extending the sight of seafarers at sea to accomplish traditional navigation functions. However, the IMO regulatory framework has failed to keep up with new technologies and instruments that can also enhance safety of navigation on conventional ships. With the advent of MASS, new sensor capabilities that can be applied to both manned and autonomous ships are being considered. Several of these extend the functionality of existing systems by providing new features, while others provide entirely new abilities that have not been possible in the past. A discussion of instrument characteristics, the types of data they can produce and their application to enhance vessel situational awareness is provided in the paragraphs that follow.

#### 6.3.2.1 Surface Sensors and Systems

Augmentation of present IMO-mandated vessel environmental sensor systems with further capability is essential to achieving comprehensive situational awareness for MASS and to ensure proper supervision and traceability of decision making. These sensor systems can expand upon existing capabilities and can also provide new capabilities not presently available, which, through the fusion of diverse data sources, can provide unprecedented levels of vessel situational awareness. Further, the integration of shipboard sensor data with external data and information resources available from space-based sensors through broadband communication channels provide the fundamental building blocks for cooperative decision making between vessels and shoreside operators, and locally between vessels using wide area network (WAN) established among the vessels themselves. Several of these systems are described below.

*Radar* – Radar supplements the human senses to help detect and avoid other vessels, to view ATON and land masses by using radio waves to detect objects and determine their range, angle, velocity and, in some cases, their characteristics. Analog signals requiring high levels of power traditionally used to implement this technology have been expanded to include digital signals that require far less power, which can be modified and adjusted to interrogate targets with different methods to extract additional information not possible using traditional analog techniques. Radar possesses the ability to produce velocity data regarding an object through use of the Doppler effect that senses changes in frequency as a result of movement in relation to the Radar transmitter. Automatic Radar Plotting Aid (ARPA) features can provide Radar contact object plotting to track course, speed, closest point of approach (CPA) and time of CPA as a means to determine the danger of collision with other ships, land masses or other objects.

*Electronic Chart Display Information System (ECDIS)* – ECDIS is a navigation computing system that displays electronic navigation chart (ENC) information including soundings resulting from hydrographic surveys of the areas sailed, locations of channels and ATON, and known hazards to navigation that may be encountered along the route. These systems comply with IMO regulations regarding the use of electronic charts as an alternative to paper navigation charts and allow for frequent and comprehensive updates both shoreside and while en route.

*Automated Identification System (AIS)* – AIS is a transponder-based tracking system used for maritime vessel and aircraft (such as search and rescue) activities including vessel traffic services. This system uses VHF radio signals with range to approximately line of sight (15–25 nautical miles) to supplement Radar as a means to avoid collision at sea and also provides a wealth of information on nearby vessels related to position, speed and identity as well as ports of origin and destination. These signals are also received by a growing number of satellites that can provide worldwide reception beyond the capabilities of shore receiving stations.

*Video and Infrared (IR) Cameras* – The use of high-resolution daylight, low light and infrared video cameras to provide 360-degree visibility of the ship's environment in a ship is a key element in moving remote controlled and automated ships toward realization. Without seafarers on board to perform watchkeeping duties and identification of physical aids to navigation, video monitoring must provide shoreside operators and onboard reasoning and decision-making capabilities with unobstructed views to enhance situational awareness both internal and external to the vessel. Rather than merely providing a video stream, many camera systems contain artificial intelligence technology that, possibly through cooperation with other sensor systems, enables the camera itself to identify objects and events of significance. A large part of video surveillance also helps mitigate security vulnerabilities [Earls 2016]. This includes the capability to continuously monitor critical areas on the vessel and the surrounding waters [ISPS 2004; MTSA 2002].

*Audio and Sound* – Vessel sound monitoring has already been considered by IMO as regards noise levels within machinery, navigation, accommodation and other normally occupied spaces [MSC.337(91)]. Although this resolution is intended to reduce risk to human safety, the methods involved provide guidance for the installation of acoustical monitoring devices for monitoring and detecting failures of critical machinery within remotely operated and autonomous vessels. Sound monitoring can also be extended to areas surrounding a vessel to fulfill the hearing requirement of COLREG Rule 5 (Proper Look-out) to detect audible ATON (bells, gongs, foghorns, etc.) as well as vessel sound signals while operating in restricted visibility. Similar to camera monitoring, acoustical monitors (microphones) can be distributed to ensure 360-degree coverage around the perimeter of a vessel to detect sounds from any location. Localization of sound sources can be accomplished using beam-forming techniques based upon the distribution of signal strength around the vessel [Papez and Vleck 2016]. The use of statistical and AI-based data analytics can detect and identify specific sounds indicative of pending or actual machinery failure, the human voices of terrorists, breaking waves on shoals, vessel and ATON sound signals and other sounds significant to vessel operations and useful to notify remote operators and onboard reasoning capabilities of the need to take appropriate action.

*Laser Imaging (Lidar)* – Much attention has been given to the use of Light Detection and Ranging (Lidar) as a means to create a high-resolution image of the environment in applications that include terrain mapping, surveying the interiors of Egyptian pyramids, and the area surrounding a ground vehicle such as a driverless car. It operates by providing up to 150,000 beams of light per second from which detailed maps of the surrounding environment may be created. Several commercially available systems are capable of identifying details of a few centimeters at a range of more than 1,000 meters. One example of its use in maritime applications is a collaborative effort of AP Moller-Maersk (Copenhagen, Denmark) and Sea Machines Robotics (Boston, Massachusetts) to improve situational awareness, object identification and tracking [MH&L 2018]. Lidar also forms the basis for a man-overboard detection system being developed by Velodyne (San Jose, California) and Mechaspin (Lake Mary, Florida) [Bertini 2017]. This system is described as capable of detecting and measuring any object that falls from a ship, making appropriate notifications, and can assist in locating that object in the sea. Such capability would help fulfill MASS duties to perform rescue and provide aid at sea. Lidar performance is decreased in precipitation such as rain, snow and fog.

*Millimeter Radar (mmRadar)* – While providing lower resolution than comparable Lidar systems, the use of mmRadar systems operating at frequencies between 75 and 100 GHz for close navigation in all weather conditions is presently routinely accomplished by military helicopters. It is also the primary forward-looking sensor used in Tesla automobiles. An advantage of mmRadar over Lidar is the ability to measure relative speed and velocity of moving objects using Doppler frequency shift. With existing naval vessel installations for defensive purposes already in use, mmRadar is also making headway as a means to reduce risk of vessel collision [Nanoradar 2019].

*Inertial Navigation Systems (INS)* – The use of INS is approved but not mandated by the IMO as a ship's primary heading indicator to meet the performance standards required by gyrocompasses (i.e., ISO 16328, 60945 and IEC 61162) and to provide data to Radar, ECDIS and voyage data recorders. The types of navigation tasks performed using INS include route planning, track control and collision avoidance. Such systems can provide much-needed backup capability to help overcome and recover from GNSS outages and spoofing events. INS can provide insight into ship movement characteristics in a seaway to detect, distinguish between, and identify the unique signatures of a change of center of gravity due to shift of load or ballast; vibration resulting from damage to ship's equipment; and abrupt actions due to lurch, sway or other effect of collision or allision. It can also detect and identify changes in wave patterns resulting from advancing frontal activity and shifts in weather systems that may be useful in comparison with forecast conditions.

*Enhanced Loran (eLoran)* – An updated version of radio navigation (eLoran) is presently being evaluated that is based upon new technology to enhance Long Range Navigation (Loran), the predecessor of GNSS. The use of widely spaced terrestrial radio transmitters to supply powerful low-frequency signals provides precision navigation and timing (PNT) data that is much harder to jam and spoof than the low power satellite signals of GNSS. South Korea is expected to have three active eLoran beacons by 2019 [Gallagher 2017]. With the signing of the National Timing Security and Resilience Act, the United States is tasked with establishing a terrestrial backup timing system for GPS by 2020 [NTSRA 2018].

*Meteorological Instruments* – A basic complement of weather instruments has been onboard vessels for decades, if not centuries, and generally include real-time data on wind speed and direction, temperature, barometric pressure and humidity. Sea temperature is also readily available from modern echosounders. Such sensors are vital for onboard reasoning capability detection and compensation for the effects of wind, currents and other meteorological and oceanographic (METOC) phenomena on MASS performance throughout the duration of a voyage. Direct integration of meteorological instruments into the overall vessel sensor fusion architecture is essential to accomplish this task.

*Remote and Autonomous Aerial Vehicles (RAV/AAV)* – The use of drone remote and autonomous aerial vehicle (UAV) (RAV/AAV) aircraft is gaining acceptance across the maritime community for security, environmental surveillance, search and rescue, and ship and cargo inspection [Vella 2018]. When adapted to remote and autonomous vessel operations, with appropriate sensors and integrated into the onboard sensor fusion architecture, they can be used very effectively for many purposes. These include serving as a lookout ahead of a vessel to detect and/or confirm traffic, hazards and threats; to inspect the hull above the waterline, deck and ship's machinery; to assess safety and environment considerations in the event of accidents and mishaps; and to detect and identify survivors during search and rescue operations.

### 6.3.2.2 Subsea Sensors

Seafarers develop skills and techniques over years of experience to assess changes to the environment that can indicate hazardous sea states and bottom conditions that compromise safety of navigation. Visual clues include changes in sea color during an approach toward a shoal, water temperature change, and breaking waves or areas of calm among rough seas without obvious cause. Without anything more than an echosounder to provide direct information of the depth of water directly below the keel, seafarers today are expected to operate using second-hand information of the depths, hazards and obstructions along their routes of transit provided by navigation charts that may contain obsolete survey data that is years, decades or even centuries old [Wright and Baldauf 2016b]. A significant number of navigation charts lack soundings altogether, especially in the Arctic, South Pacific, Indian and the South Atlantic Oceans and other parts of the world that are poorly surveyed or not surveyed at all. MASS operations must compensate for lack of human knowledge and expertise as well as deficient charts by providing sensor capabilities to directly assess bottom configurations and conditions in real time. Sensor systems that can provide these capabilities are described below.

*Echosounder* – Single-beam echosounders mandated as standard equipment carriage requirements by the IMO traditionally provide a 1- to 4-digit representation of depth below the keel measured in units such as feet, meters or fathoms. Modern echosounders now also provide sea temperatures both at and below the surface; bottom contours along the route of transit; and bottom consistency including mud, sand and rock, and other characteristics of the seabed.

*Navigation Sonar* – Sonars that provide look-ahead capability in the direction of transit can significantly enhance safety of navigation to detect hazards, shoals and changes in bottom contours occurring since the last hydrographic survey due to storms and currents [Wright and Russell 2017]. These Sonars have become essential for use by expeditionary vessels, luxury yachts and ocean liners that travel in poorly charted areas of the world.

Several manufacturers produce Sonar products that provide forward-looking capabilities with range of approximately one hundred meters. These systems are generally adequate for small vessels that are agile and can maneuver and stop quickly, but they are not well suited to vessels that need much greater advance notice of hazards to alter course or cease forward motion. Systems with forward range of 500–1,000 meters or more are much more suited to larger vessels and can be readily integrated into a remotely operated or autonomous vessel's sensor fusion architecture to facilitate artificial intelligence-based decision making and navigation. One manufacturer of a forward-looking multibeam Sonar (FarSounder, Inc, Warwick, Rhode Island) introduced new capabilities to supplement charts with high-resolution swath bathymetry captured along the route of transit for use in future navigation chart development [FarSounder 2017]. They also contribute Sonar data to the U.S. National Oceanic and Atmospheric Administration (NOAA) as a trusted node under International Hydrographic Organization (IHO) jurisdiction through the Crowdsourced Bathymetry Working Group (CSBWG) [FarSounder 2018]. The ability of navigation Sonar to enhance safety of navigation for all vessels, especially in the Arctic and other remote areas, should be considered by the IMO in the update to the Polar Code and for amending vessel carriage requirements in general. [Wright, R. Glenn and Michael Baldauf 2016d]. Mandating navigation Sonar use for MASS will provide layers of redundancy for detecting ATON, other vessels based upon their below-the-waterline profile; hazards to navigation such as icebergs, growlers, shipping containers adrift and large marine mammals; and bottom hazards such as reefs and shoals.

*Side-Scan Sonar* – Self-contained, individual side-scan Sonar systems and even relatively inexpensive single-beam echosounder systems now include side-scan Sonar capability that provides vivid and high-resolution imagery of the bottom and bottom features such as rocks, wrecks, and even water column contents such as fish. These new capabilities provide increased utility to supplement GNSS and INS positioning data. This may be accomplished using contour following across the seabed as well as object recognition using AI techniques to detect and identify features and landmarks that are prominent on the bottom and calculating position based upon their location [Wright and Baldauf 2016c; Wright 2019a].

*Remote and Autonomous Undersea Vehicles (RUV/AUV)* – In a manner similar to RAV and AAV, undersea vehicles (AUV) can be adapted to remote and autonomous vessel operations to look ahead and identify hazards to navigation as well as to perform underwater hull surveys and examine propellers and rudders for damage and obstructions that hinder their operation. Much research has been performed to integrate a wide variety of sensors into these vehicles as well as AI-based object detection, avoidance and navigation capabilities. Acoustic transducers provide adequate bandwidth for nearby wireless imagery transmissions to the vessel suitable for real-time decision making and working in tandem with MASS.

*Security Sonar* – The detection, tracking and classification of divers as well as underwater vehicles that may pose threats may be accomplished using security Sonar. Such systems are available from a variety of manufacturers and their ranges vary from a few hundred to a thousand or so meters to create a secure perimeter through 360 degrees around a vessel. Weaknesses may exist in their ability to detect divers using closed-circuit systems such as rebreathers, and to distinguish between divers or underwater vehicles and larger marine mammals, fish and schools of fish. Enhancements to existing algorithms using artificial intelligence and other methods to better distinguish between different targets based upon their unique signatures expressed in Sonar signals have often proved successful in enhancing the performance of these Sonar systems.

### 6.3.3 Air-Based Sensors

The use of aircraft as well as remotely controlled and unmanned aerial vehicles (RAV/UAV) to act as platforms for a wide variety of visual, infrared, Radar and other sensors has provided over-the-horizon visibility of vessel traffic, reef and bottom configurations in shallow waters and hazards to navigation such as icebergs and growlers. Communications can be achieved using conventional radio signals or via satellite links, depending on the physical distance between the aircraft and the vessel as well as the resolution of the imagery and sensor data required to be available for analysis.

### 6.3.4 Space-Based Sensors

As of 2018 there were approximately 4,600 satellites in Earth orbit, of which nearly 2,000 were operational [UNOOSA 2018]. One report shows the growth in satellite launches increasing three-fold over the next decade with 3,323 satellites with a mass over 50 kg launched and to be launched between 2018–2027, as compared to 1,019 satellites launched between 2008 and 2017 [Satnews 2019]. Many of these satellites are used in maritime operations to gather meteorological, oceanographic and terrestrial imaging and datasets, as well as providing GNSS positioning and timing information. However, much of the increase in satellite launches represent a new generation of small satellites sent to low earth orbit to create constellations of thousands that will provide ubiquitous global broadband access. This trend has already been noted with the announcement by Inmarsat that their Fleet Xpress service launched in March 2016 had by early 2017 passed the 10,000-ship milestone [gCaptain 2017]. The market for broadband connectivity in the maritime sector alone is expected to generate $4.7 billion by 2027 [NSR 2018]. Broadband satellite connectivity is essential for monitoring MASS operations; for sharing large volumes of sensor imagery, data and results of onboard decision-making processes; and for facilitating the implementation of blockchain technology, Big Data and Internet of Things (IoT) applications to ensure safe and secure operations.

#### 6.3.4.1 Global Navigation Satellite System (GNSS)

The primary source of precision navigation and timing (PNT) information for shipping is provided by one or more of the GPS, Galileo, GLONASS, BeiDou and regional satellite systems that comprise GNSS. This fact remains unchanged with the introduction of MASS. However, the reliance of onboard reasoning capabilities on the use of GNSS to correlate real-time, precision imagery and data in the creation of location-enabled intelligence is critical to ensure situational awareness. Higher degrees of accuracy are needed for inland and port navigation, where maneuvering is required in close proximity to fixed shore infrastructure and to ultimately secure the vessel quayside, than is generally required to transit open bodies of water. However, high overall accuracy is needed to accurately estimate range, bearing, closest point of approach and timing of vessel traffic, hazards to navigation, land masses and undersea features.

Although GNSS is heavily relied upon as the primary source for PNT data, suitable technologies are needed to supplement its use and to provide backup capability in the event GNSS services become unavailable.

### 6.3.4.2 Automatic Identification System (AIS)

While required shipboard AIS equipment is useful for gaining situational awareness of vessels within VHF radio transmission distance, maritime domain awareness of all vessels pertinent to MASS voyages as a whole regardless of their locations can be obtained from satellite-based AIS tracking. Such information can be useful for long-range tracking of other vessels relevant to MASS voyages that are outside the range of VHF radio-based AIS. This can provide a means for real-time remote operators and onboard reasoning capability to identify support vessels, freight movements and transfers between vessels affected by delays while en route to terminals with corresponding adjustments in speed and routing to ensure most efficient passage.

### 6.3.4.3 Meteorological and Oceanographic (METOC)

METOC data is available via broadband satellite connections while a ship is outside the range of terrestrial data communications. METOC data is also available while in and nearby ports using Wi-Fi and during a large portion of near-coastal voyaging while in receiving range of broadband cellular links. Onboard reasoning capabilities can make use of these data for voyage planning prior to departure, to update planning while underway based upon the latest reports, and to effect safe arrival at the destination. This includes real-time conditions from buoys and stations as may be available for tides and currents, wind, humidity and temperature. Also available is visual, infrared and water vapor imagery of cloud cover. Numerical data includes wind and swell height and direction and sea state conditions; air/sea temperatures and barometric pressures; sea ice and other physical phenomena; surface visibility; tropical cyclone movements and forecasts; and a great deal more information useful for MASS operations.

#### 6.3.4.4 Other Sensors

Other sensor inputs that may be considered to assist in MASS operations include space-based synthetic aperture Radar to detect surface features that may include other vessels or hazards to navigation, shipboard synthetic aperture Sonar to detect seafloor features for navigation, and Long Range Identification and Tracking (LRIT) system vessel positioning information.

### 6.3.5 Sensor Data Types and Characteristics

The various types of data that may be obtained from maritime sensors can be categorized according to the *pixel*, *time* and *frequency* domains that represent different perspectives of the environment. The *pixel* domain represents a translation of a spatial quantity into a pixel representation by capturing an image of a scene or object directly onto picture elements, or pixels, each of which contains an impression of the qualities of a small portion of the overall image. The original scene or object is reconstructed by means of reproducing the pixel impressions onto a display. This is the case for digital and infrared cameras and other visual sensors. Changes in imagery that occur as a function of time are reflected in the *time* domain. Different mathematical and statistical functions can be applied to pixel and time domain representations to extract data and correlate information regarding image content. Direct to pixel domain imagery is limited based upon the size and resolution of the sensor and can be enhanced using optical magnification and greater numbers of smaller pixels, as well as through the use of image filtering and software analytics.

Radar, Sonar and Lidar images are created using an entirely different process involving one or more transducers (antennas) that transmit and/or receive signals. These signals are analyzed in the *frequency* and time domains and subsequently converted into the different domain representations. A simplified example is provided in Figure 6.2, where received waveforms are analyzed in the frequency and time domains (a and b) to create a pixel domain representation (c).

Highly complex waveforms across many frequencies are projected from a transducer/antenna onto a physical scene or landscape that modifies the waveforms through reflection and absorption based upon the physical and electrical characteristics of the objects within the scene. A portion of the transmitted signals are reflected back to and received by the transducer that are analyzed as a function of changes that occur over time as well as changes detected in the frequency of the signal. Analysis of the resulting Radar and Sonar signals in the time and frequency domains is performed to acquire the information necessary to subsequently create a pixel domain image for display to seafarers. Lidar and mmRadar displays for maritime use are still under development and are not expected to be used in conventional shipping in the foreseeable future.

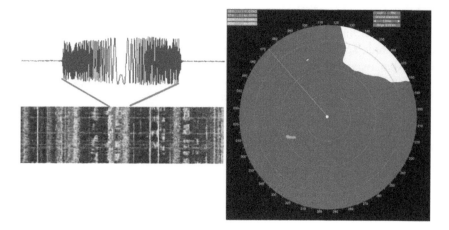

*Figure 6.2* Chirp waveform variation over time, with the resulting pixel domain representation of the local environment. (a) Radar chirp waveform showing change in frequency. (b). Waterfall representation of waveform variation over time. (c). Pixel domain representation of processed waveforms.

While this indirect approach has proven to be highly accurate and reliable, it can result in a great deal of variability in how the targets and scene are displayed to the user based upon signal resolution and manufacturer user interface design preferences. A target may be represented as a "blip" on a Radar screen, and navigation Sonar can paint a surface model of bottom terrain, while Lidar and mmRadar systems can display a highly accurate model of the terrain and quayside environment. Unlike imaging sensors, the information contained within the received signal in the time and frequency domains used to create the resulting pixel domain image is based upon the properties of the waveforms being transmitted, the gain and resolution of the transducer and antenna elements, the sensitivity of the receiver and the capabilities of the software to analyze the reflected signals.

The ability to actively interrogate targets using a wide range of waveforms provides greater flexibility to analyze their reflected signal properties across all three data classes in the pixel, time and frequency domains. The ability to dynamically adjust waveform signal characteristics in real time based upon target properties, and to further mine existing time and frequency domain datasets, continues to result in the retrieval of much greater information content than was previously possible. Recent examples in the case of Sonar data include the acquisition of swath bathymetry from navigation Sonar and other scientific data from high-resolution side-scan Sonar imagery [FarSounder 2018; Wright 2017]. Similar advances have also occurred in other maritime applications that include improvements in solid state Doppler Radars.

## 6.3.6 Sensor System Limitations and Vulnerabilities

Significant limitations exist with key sensor technologies and the data infrastructure upon which MASS must rely for safe and proper operation. In addition, vulnerabilities also exist in these same technologies that, even when they operate correctly as designed, make them susceptible to outside influences that can degrade or interrupt their proper operation. Multiple sensor modalities to provide complementary and redundant capabilities must be considered to guard against single point of failure scenarios where the loss of even one sensor can have catastrophic results on life, property and the environment. Likewise, significant effort is needed within the data pipeline and storage infrastructure to protect against hacking and piracy to ensure the security and validity of sensor data.

Specific details are given for GNSS and AIS vulnerabilities and their potential effects that can jeopardize MASS operations. Also described are examples of how hacking may affect ATON, navigation chart and other databases that can adversely affect ECDIS and electronic charting systems. Further discussion is provided to illustrate the benefits of redundancy through the use of multiple, complementary sensor technologies. These examples are not meant to be comprehensive, but illustrative of the types of vulnerabilities to which all ships, and especially MASS, are susceptible.

### 6.3.6.1 GNSS Outages, Spoofing, Jamming and Denial of Service

The reliance on GNSS by MASS onboard reasoning capabilities to correlate imagery and data to ensure comprehensive situational awareness is made even more critical as a result of the denial of service and spoofing attacks, effectively making this a single point of failure condition that may have disastrous results.

GNSS signal outages have been known to occur where complete system failure has resulted in the unavailability of services [Gibbons 2014]. This issue has become less of a problem with the advent of multiple GPS, Galileo, GLONASS, BeiDou and other constellations from which position calculations may be made. However, all GNSS signals are subject to being jammed and spoofed to interrupt their reception entirely, degrade their performance and reliability, or even to deceive as to a vessel's exact location. Such events can result in vessel groundings or redirection into hostile waters, making them vulnerable to hijacking [Forssell 2009]. This capability was demonstrated by researchers at the University of Texas at Austin, where a yacht was driven well off course and was essentially hijacked using spoofing techniques [Zaragoza 2013]. This was also reported by the U.S. Coast Guard acknowledging this phenomena as being of concern beyond the maritime industry and to include the transportation sector as a whole [Thompson 2014]. These techniques may also be used against MASS and MASS-based UAVs in much the same manner that Iran reportedly used to capture a U.S. Central Intelligence Agency RQ-170 Sentinal UAV by jamming its communications and reconfiguring its GPS coordinates to land in Iran rather than at the intended destination in Afghanistan [Peterson and Faramarzi 2011].

The jamming of GNSS signals can have the same effects as an outage was demonstrated in 2010 when numerous, low power personal privacy jammers were detected as interfering with GPS involving airport operations at Newark, New Jersey [Grabowski 2012]. Many such problems have been widely reported over the past several years near Korea, in the eastern Mediterranean and Red Seas, and in the western Baltic Sea. Political events can also spur long-term interruptions in GNSS service as evidenced by British plans to develop an alternative satellite system to Galileo [Gov.uk 2018].

### 6.3.6.2 AIS Range, Clutter, Spoofing and Jamming

AIS broadcast range is normally limited to the line of sight and is subject to the effects of atmospheric propagation where maximum reliable ship-to-ship radio communications over sea water is in the range of 20–25 nautical miles. Shore stations, with high antennas, can reliably receive AIS messages from ships at distances of up to 20–35 nautical miles [ITU 2007]. While this range is more than adequate for determining which traffic may be of concern in collision avoidance, the information displayed on an AIS receiver may not accurately reflect reality in terms of whether the vessels and their characteristics are properly listed and are present at their indicated locations. The probable causes of problems with AIS have been identified by the International Association of Marine Aids to Navigation and Lighthouse Authorities (IALA) as including incorrect data input to AIS unit, disruption to GNSS, failure of AIS unit, degradation of VHF propagation, loss of VHF reception and control system malfunction [IALA PAP26-8.1.3]. IALA concluded that AIS does not have inherent integrity or authentication, that it is possible to broadcast false information via AIS, that AIS should not be used as a sole source of information, that other means should be used to verify AIS information and that the integrity of AIS information should be monitored.

Other problems with AIS that reflect on limitations of the technology used in its implementation have been reported. For example, reports have identified the inappropriate use of AIS tracking buoys to mark fishing nets [Kovary 2018]. In addition to an inability to determine whether the indicated target is an actual vessel or a fishing net, the sheer number of AIS transponder buoys has overwhelmed AIS receivers with too many targets to simultaneously process, resulting in clutter with hundreds of targets in one area and no targets in others.

The ability to spoof and jam AIS broadcasts has particular significance where AIS-ATON signals are used for vessel navigation. A lack of security controls can facilitate a ship being diverted off course by placing AIS-ATON in undesirable or even dangerous locations inadvertently, for hijacking or for other nefarious purposes [Simonite 2013]. The vulnerabilities of AIS have also resulted in its use by criminals to attempt to evade law enforcement [Middleton 2014]. Another report found that AIS data is being increasingly manipulated by ships that seek to conceal their identity, location or destination for economic gain or to sail anomalously and without AIS use. The report concludes that this is a fast-growing, global trend undermining decision makers who rely, unknowingly and unwittingly, on inaccurate and increasingly manipulated data [Windward 2014].

### 6.3.6.3 Database Hacking

One of the greatest dangers associated with ECDIS and the electronic navigation charts (ENC) they display is their primary existence as a collection of data objects in cyberspace without having a traditional physical presence to provide backup in the event of their electronic corruption or disappearance. This property makes them susceptible to hacking and cyber attacks that can render them useless or even detrimental and hazardous to navigation. Widespread corruption can occur at the source databases within which ATON and AIS-ATON objects, soundings, hazards to navigation and other data reside at the authorized service provider. In the United States this responsibility is shared between the Coast Guard for ATON and the Light List, and the National Oceanographic and Atmospheric Administration (NOAA) for ENC. Corruption can also occur at the local level where individual or groups of ATON, soundings and other data in the same geographical area may be corrupted through directed attacks at individual vessels or fleets accomplished through updates to corrupted database products as well as by viruses and other malicious software (malware) installed on the host computers on the vessels themselves.

Initiatives exist at both the U.S. Coast Guard and NOAA aimed at defending their computer networks from attacks to prevent such occurrences [Radgowski et al., 2014; NOAA 2014]. Both initiatives acknowledge the threats involved and are steps in the correct direction to manage and even overcome the adverse effects on national security imposed by these threats. Issues that pertain to ATON and ENC design, development and implementation cross agency lines, barriers and firewalls, making the solution to these problems even more difficult. Comprehensive security measures on the part of fleet and vessel operators are also needed to restrict physical and on-line access to ECDIS and other vessel systems, to ensure that the installation and download of ENC is accomplished safely and that frequently updated virus and malware detection software is used to detect and render malicious code ineffective within onboard computers.

### 6.3.6.4 Multiple Sensor Modalities

Absent on-scene availability of human senses and knowledge during MASS operations the duplication of critical sensor capabilities is vital to increase the safety of navigation and to ensure the vessel can continue operations without interruption regardless of the mode of transport. A recent example by Boeing in their 737 MAX 8 design is the use of a single sensor to measure angle of attack as part of their Maneuvering Characteristics Augmentation System (MCAS) – a design flaw theorized to be a possible cause of the 29 October 2018 Lion Air crash in Indonesia and the 9 March 2019 Ethiopian Airlines crash in Africa [Tangel and Pasztor 2019]. However, the best solutions include not only multiple sensors of the same type, but complementary sensors that use different technologies to achieve the same result.

A common example is the use of INS to backup GNSS as a means to determine PNT, where multiple instruments of each type can provide clues as to the failure of one device using methods such as a voting scheme such as between three independent INS and/or GNSS instruments. However, in the event of a GNSS spoofing or denial of service attack this method would not be effective since all GNSS would likely differ from all INS. Further complicating the matter is the fact that the periodic update of INS positioning using GNSS to overcome integration drift due to accumulating acceleration and angular velocity errors would not be possible. One method to break this deadlock would be the availability of a third means of position determination using ATON by which a fix can be made to determine position, provided the ATON has been recently verified as watching properly in good waters. Another method would be to reference an eLoran fix or to implement georeferencing using bottom contour tracking with an echosounder as a third method for position determination. Several such methods are available, but increasing levels of complexity and uncertainty associated with the proper integration of these modalities greatly complicate matters. A similar argument can be made for supplementing the required single-beam echosounder with a second echosounder, each with independent transducers. Such redundancy is prudent but may result in dissimilar measurements due to differences in hull installation locations of the separate transducers and the use of different frequencies to ensure interference between instruments does not exist. A better solution would be to supplement a single-beam echosounder with forward-looking navigation Sonar that provides far greater swath coverage of the bottom, which can be used to check the validity of the single-beam echosounder data.

The use of multiple and complementary sensor modalities is essential to achieve comprehensive situational awareness during MASS operations. Many of the sensors described are not presently included in IMO equipment carriage requirements, but should be in the future to ensure the greatest possible safety margins not only for normal operations, but also for when situations do not go according to plan to provide greater assurance of safe resolution and recovery in the absence of direct human supervision.

## 6.4 NAVIGATIONAL REASONING

Previous paragraphs described the different types of aids to be used in MASS navigation; the sensors available to detect, identify and use these aids; and the types of signals that are available from these sensors for subsequent examination by intelligent navigation reasoning capabilities in determining the proper course(s) of action to be taken. This last section describes the technologies that are available to acquire the sensor data and signals and to reason with their content to reach the appropriate conclusions in a manner that is logical and verifiable. A minimum goal for such reasoning systems is to achieve a level of confidence that is at least equal to human performance. However, humans make mistakes and a higher goal must be set not only to avoid many of the mistakes made by humans but also to achieve greater results than are humanly possible. Machine learning and deep-learning artificial intelligence (AI) technologies form the basis to integrate multiple sensor modalities into a cohesive approach and to accomplish decision making for MASS navigation. Considering the scope of the data and imagery available from onboard and space-based sensor systems described in Section 6.3, this paragraph takes a look at the processes and methods used by AI-based navigation products.

Many tasks are performed to accomplish the complex operations associated with navigating MASS from a point of origin and along a route of transit to arrive safely at a destination. The principal task of selecting an appropriate route may seem rather trivial in today's world, where we merely select the destination for our vehicle and a navigation system plans the route based upon our present location; examines known traffic, construction and other delays; and calculates an estimated time of arrival based upon an expected rate of progression along the route. However, as anyone who has been led through parking lots, down dead-end roads, directed across non-existent bridges and told to exit the highway in the middle of a field knows – it is often necessary to adjust and modify our strategy to proceed to a destination when unexpected conditions are encountered while en route. Thus it is vital, especially in the case of MASS, for intelligent navigation systems to be capable of independently reasoning by planning and structuring information so that small to large portions of their strategies can be reorganized quickly and precisely based upon real-time events. Such capabilities are realized through the implementation of distributed processing where smart sensors capable of detecting and identifying objects and events provide this information directly to the navigation system that can rapidly assess the information provided to make the appropriate decisions. This process is described in the following four general scenarios for discussion purposes. Actual relationships between multiple integrated sensors and the intelligent navigation reasoning system have been simplified and these scenarios are not intended to represent a comprehensive depiction of the capabilities of these technologies. Also, since MASS equipment carriage requirements are still being determined in the second phase of the IMO regulatory scoping exercise, some of the instruments and technologies discussed may not be carried on individual vessels.

### 6.4.1 Navigation under Nominal Conditions

This first scenario represents navigation under nominal conditions where positive position determination is continuously maintained and progression along a proper route of transit is accomplished using an electronic adaptation of navigation by pilotage. Radar, AIS and smart sensors in the form of visual and infrared cameras detect buoys, natural landmarks and other ATON, apply their training to discern their unique identities based upon their indicated characteristics, and pass this information to the navigation system along with all other reporting sensors. ATON position fixes are established by the navigation system based upon vessel position and relative bearing from the sensor determined using precise timing and bearing calculations considering lateral offset and direction of the sensor location from the GNSS antenna. Comparison is then made with ECDIS and the ENC database to determine whether the sighted ATON corresponds to the identity and watch circle position of the expected ATON. Once successfully accomplished, the navigation system determines the next ATON in logical sequence, its expected position and the relative bearings for each sensor where it is expected to be in view, and the anticipated time it is to be encountered. This process is repeated on a continuing basis throughout the voyage.

Simultaneous with ATON detection and identification, other surface sensors are monitored including INS and eLoran (if available) positioning for comparison with GNSS positioning as being within expected tolerances. Radar and, when the vessel is properly equipped and within sensor range under the appropriate weather conditions, Lidar and mmRadar imaging of the environment can be used to assess expected versus actual, detected land features. Meteorological instruments, tides and currents are monitored to determine whether course offsets are needed to overcome windage and other forces external to the vessel and to predict the potential for future problems based upon channel widths and other factors. Radar and AIS are also used to detect other vessels and to determine whether their paths may conflict with the vessels' own path.

Subsurface sensors are monitored simultaneously with surface sensors to ensure comprehensive situational awareness is maintained. The single-beam echosounder traces a line of soundings across the bottom to ensure safe margins of depth exist below the keel and to verify the vessel in on the proper route of transit when compared with soundings contained within ENC hosted in ECDIS or other chart viewing equipment. Navigation Sonar with forward-looking capabilities across a swath ahead of the vessels' route of transit can also be compared with the corresponding area on the ENC and to detect shoal waters, rocks and reefs, sunken vessels and other hazards to navigation as well as large marine mammals that should be avoided. Side-scan Sonar imagery used to detect unique bottom features can supplement navigation by pilotage using surface ATON and prominent landscape elements. AI and machine learning statistical analytics capabilities are used to examine subsea sensor data and imagery, recognize and identify significant trends and features, and detect deviations from the expected.

### 6.4.2 Navigation Absent Expected ATON

Many different scenarios exist where ambiguity and uncertainty over actual vessel position may be in question. This can be caused as a result of failure in sensor systems used to detect physical phenomena and the movement, loss damage or destruction of physical and AIS-ATON due to storm action, allision or movement of ice. In the case of MASS this may also be caused by Radar and forward-looking navigation Sonar failure to detect ATON presence, and by smart visible and infrared imaging sensor failure to detect specific ATON characteristics to correctly identify specific aids, or incorrectly identify an ATON as another aid. Another possibility is that an ATON is off its assigned position or missing entirely from the ENC.

In each of these cases, an intelligent navigation reasoning system will analyze whatever built-in test capabilities may be available for each sensor system to determine whether these systems are faulty. Absent any specific fault information, the sensor data and imagery will be examined from the various distributed sensors tasked to detect ATON and correlate sensor and image information (or lack thereof) with positioning data provided by GNSS, INS, eLoran, and Sonar systems as may be available. Multiple redundant sensor systems using different technologies and processes to arrive at ATON identification and positioning information, for example, can greatly increase the probability of arriving at the correct conclusions needed to resolve ambiguities, the problems encountered and to maintain voyage progress. Failure to detect and identify several ATON in direct succession becomes problematic but not necessarily limiting as long as correct positioning appears to be maintained using GNSS as well as sources of positioning information that are immune from denial of service and spoofing attacks. Exact performance limits for tolerating ambiguities and anomalies in sensor information must be determined in accordance with voyage safety plans specific to the routes and the vessel.

### 6.4.3 Navigation under Threat of Collision, Allision or Attack

An intelligent automated navigation system would work under the assumption that a constant threat of collision, allision or attack exists with all Radar, AIS, visible and infrared contacts unless they are well passed and proceeding in a direction opposite the vessel, and at a distance and rate of speed where a rapid change in course would still provide an adequate timeframe to properly respond. This assumption would require the MASS vessel to constantly monitor and plot corrective and/or evasive actions that are not to be put into effect until specific and graduated limits in terms of CPA and time of CPA are exceeded. A secondary assumption is that the actions of all contacts and the MASS vessel itself in responding to contact actions will proceed according to the COLREG convention. A third assumption is that when contacts violate the COLREG Convention by their actions the MASS vessel should still abide by the COLREG Convention unless it is no longer possible to do so, after which any and all actions required to react to and avoid the contact would be permitted.

Such a scenario requires the implementation of an intelligent automated navigation system capable of learning about the proper behavior of vessels in closing situations both according to the COLREG Convention and in actual practice. This approach exhibits the characteristics of machine intelligence through directed learning by experiencing proper reaction and avoidance techniques fine-tuned and adjusted based upon human behavior under such circumstances. Learning can be further reinforced using deep-learning AI techniques that take advantage of literally millions of Radar, AIS, visible and IR images as well as Sonar data and imagery viewed and analyzed during all voyages, taking into account potential threats both above and below the waterline.

### 6.4.4 Navigation under GNSS/AIS Denial of Service Attack or Spoofing

Under circumstances where GNSS and/or AIS signals and data become unavailable, unreliable or erratic as may occur during a denial of service attack, or are consistently diverging from alternative positioning technologies such as inertial navigation, eLoran and/or single-beam echosounder ground track monitoring as may occur during a spoofing attack, it is crucial that the affected instruments be identified, isolated and removed from service as quickly as possible and, if conditions permit, the voyage continued using alternative positioning methods. Such conditions may exist momentarily over a few minutes, temporarily over several hours, or be of longer duration in terms of days or weeks. Much as an intelligent automated navigation system was trained on the use of the COLREG for collision and allision avoidance, the same methods should be used to train the system to operate using different combinations of positioning sensors simulating service outages under various conditions.

### 6.4.5 Digital Twin Simulation

The starting point for all training of intelligent automated navigation systems should be accomplished using digital twin simulation with actual sensor data recorded live over the same routes planned for use by MASS as well as alternative routes that may be used should the need arise. Once initially trained, additional learning should continue while en route under actual cruise conditions with provisions for integrating new knowledge and experience using well-controlled procedures on a reoccurring basis. Such training should be accomplished continuously throughout the entire lifecycle of the MASS project until sufficient experience validated by appropriate metrics indicates all aspects of the system are mature. Each of the three aspects of the system may not mature at the same rate. For example, COLREG compliance training may mature far faster than collision, allision and threat avoidance. The same can be said for learning using partial sensor system availability, where the various combinations of partial sensor data and information may require years of training using millions of sets of sample data.

### REFERENCES

Bertini, F. 2017. Cruise Ship Overboard Detection with LiDAR. Velodyne, 12 July 2017. https://velodynelidar.com/newsroom/cruise-ship-overboard-detection-lidar/.

COLREG 1972a Rule 5. International Regulations for Preventing Collisions at Sea. Rule 5 Look-out.

COLREG 1972b Rule 7. International Regulations for Preventing Collisions at Sea. Rule 7 Risk of collision, (b) and (c).

Earls, A. R. 2016. From an Interview, 'More Shipowners Use Video Cameras for Security Safety and Operations'. Professional Mariner, June–July 2016, 27 May 2016. http://www.professionalmariner.com/June-July-2016/More-shipowners-use-video-cameras-for-security-safety-and-operations/.

FarSounder 2017. FarSounder Overlays S-57 & S-63 Charts. Press Release, 14 September 2017. www.farsounder.com.

FarSounder 2018. FarSounder Joins NOAA as a Trusted Node. Press Release, 17 October 2018. www/farsounder.com/about/press_releases.

Forssell, B. 2009. The Dangers of GPS/GNSS. *Coordinates*, February 2009, pp. 6–8.

Gallagher, Sean. 2017. Radio Navigation Set to Make Global Return as GPS Backup. *ARS Technica*, 7 August 2017. https://arstechnica.com/gadgets/2017/08/radio-navigation-set-to-make-glocal-return-as-gps-backup-because-cyber/.

*gCaptain* 2017. Connected at Sea: Inmarsat's New High-Speed Broadband Service Hits 10,000-Ship Milestone. 4 May 2017. http://gcaptain.com/fleet-xpress-exceeds-10000-ship-milestone-first-anniversary/.

Gibbons, Glen. 2014. GLONASS Suffers Temporary System wide Outage; Multi-GNSS Receiver Overcomes Problem. *Inside GNSS*, 2 April 2014. http://www.insidegnss.com/node/3979.

Gov.uk 2018. Space Sector to Benefit from Multi-million Pound Work on UK Alternative to Galileo. Press Release, 29 August 2018. https://www.gov.uk/government/news/space-sector-to-benefit-from-multi-million-pound-work-on-uk-alternative-to-galileo.

Grabowski, Joseph C. 2012. Personal Privacy Jammers: Locating Jersey PPDs Jamming GBAS Safety-of-Life Signals. *GPS World*, 1 April 2012. http://gpsworld.com/personal-privacy-jammers-12837/.

International Association of Marine Aids to Navigation and Lighthouse Authorities (IALA)PAP26-8.1.3. Draft IALA Briefing Note AIS Vulnerability. e-NAV Committee, 22 October 2013.

International Association of Marine Aids to Navigation and Lighthouse Authorities (IALA) 2014. IALA International Dictionary of Aids to Marine Navigation, cited in IALA NAVGUIDE Aids to Navigation Manual, Seventh Edition, p. 30.

International Ship and Port Facility Security (ISPS) 2004. SOLAS chapter XI-2. Part B 9.42.

ITU 2007. REPORT ITU-R M.2123, *Long Range Detection of Automatic Identification System (AIS) Messages under Various Tropospheric Propagation Conditions*, p. 32, Appendix 5 AIS Propagation Observations – Ducting.

Kovary, Laura. 2018. AIS Problems Revealed in East China Sea. *gCaptain*, 27 December 2018. https://gcaptain.com/ais-problems-revealed-in-east-china-sea/.

MAREX 2019. Autonomous Collision Avoidance Tested Using Damen Crew Boat. *Maritime Executive*, 26 March 2019. https://www.maritime-executive.com/article/autonomous-collision-avoidance-tested-using-damen-crew-boat.

Maritime Transportation Security Act (MTSA) 2002 Maritime Transportation Security Act of 2002. 33CFR 104.285 (1).

MH&L 2018. Maersk First to Test LiDAR on Container Ship. *Material Handling and Logistics*, 30 Apr 2018. https://www.mhlnews.com/technology-automation/maersk-first-test-lidar-container-ship.

Middleton, A. 2014. Hide and Seek, Managing Automatic Identification System Vulnerabilities. *Coast Guard Proceedings*, Winter 2014–2015, p. 50.

MSC.337(91). Code on Noise Levels On Board Ships. IMO Resolution MSC.337(91). Adopted on 30 November 2012.

MSC 98/20/2. Work Programme. Maritime Autonomous Surface Ships Proposal for a Regulatory Scoping Exercise. Regulatory Scoping Exercise for the Use of Maritime Autonomous Surface Ships (MASS), 27 February 2017.

Nanoradar 2019. Unmanned Boat Avoidance. www.nanoradar.cn/english/page.phd?id=30.

National Timing Resilience Act (NTSRA) 2018. National Timing Resilience Act of 2018. Signed on 4 December 2018 as Part of the Frank LoBiondo U.S. Coast Guard Authorization Act of 2018.

NOAA 2014. Office of the Chief Information Officer, IT Security Office, 15 September 2014, https://www.csp.noaa.gov/.

Northern Sky Research (NSR) 2018. Maritime SATCOM Markets, 6th edn. Press Release, 2 July 2018. https://www.nsr.com/nsr-report-full-steam-ahead-for-broadband-maritime-connectivity.

Papež, Martin and Karel Viček. 2016. Model of Surveillance System based on Sound Tracking. *Advances in Intelligent Systems and Computing*, 466: 267–78. https://doi.org/10.1007/978-3-319-33389-2_26.

Peterson, Scott and Payam Faramarzi. 2011 Iran Hijacked US Drone, Says Iranian Engineer. *The Christian Science Monitor*, 15 December 2011. https://www.csmonitor.com/World/Middle-East/2011/1215/Exclusive-Iran-hijacked-US-drone-says-Iranian-engineer.

Radgowski, Capts J. and K. Tiongson. 2014. Cyberspace-the Imminent Operational Domain. *Coast Guard Proceedings*, Winter 2014–2015, p. 18.

Satnews 2019. Euroconsult Report Focuses on Satellites To Be Built and Launched by 2027. *Satnews Daily*, 26 January 2019. http://satnews.com/story.php?number=2091711277.

Simonite, Tom. 2013. Ship Tracking Hack Makes Tankers Vanish from View. *MIT Technology Review*, 18 October 2013.

Stewart, Jack. 2018. Paraphrased from, Rolls-Royce wants to Fill the Seas with Self-Sailing Ships. *Wired*. 15 October 2018. https://www.wired.com/story/rolls-royce-autonomous-ship/.

Tangel, Andrew and Andy Pasztor. 2019. Boeing to Make Key Change in 737 MAX Cockpit Software. *Wall Street Journal*. 12 March 2019. https://www.wsj.com/articles/boeing-to-make-key-change-in-max-cockpit-software-11552413489.

Thompson, Brittany M. 2014. GPS Spoofing and Jamming-A Global Concern for All Vessels. *Coast Guard Proceedings*, Winter 2014–2015, p. 50.

United Nations Office for Outer Space Affairs (UNOOSA). 2018. Annual Report 2017, March 2018, Vienna, p. 5.

Vella, Heidi. 2018. Drones in the Deep: New Applications for Maritime UAVs. Ship Technology, 23 January 2018. https://www.ship-technology.com/features/drones-deep-new-applications-maritime-uavs.

Windward 2014. AIS Data on the High Seas: An Analysis of the Magnitude and Implications of Growing Data Manipulation at Sea. http://www.windward.eu.

Wright, R. Glenn. 2017. Scientific Data Acquisition using Navigation Sonar. IEEE/MTS *Oceans Conference*, Anchorage, AK, September 2017.

Wright, R. Glenn. 2019a. Intelligent Autonomous Ship Navigation using Multi-Sensor Modalities. 7th International Symposium on Marine Navigation and Safety of Sea Transportation (TransNav). Gdynia, Poland, 12–14 June 2019.

Wright, R. Glenn. 2019b. Enhanced MASS Situational Awareness using Virtual Navigation Aids. Second International Conference on Maritime Autonomous Surface Ships, Trondheim, Norway. November 2019.

Wright, R. Glenn and Michael Baldauf. 2016a. Arctic Environmental Preservation through Grounding Avoidance. *Sustainable Shipping in a Changing Arctic*, ed. Lawrence Hildebrand, Lawson Brigham, Tafsir Johansson. Springer International Publishing AG. ISBN 978-3-319-78424-3.

Wright, R. Glenn and Michael Baldauf. 2016b. Correlation of Virtual Aids to Navigation to the Physical Environment. *TransNav, International Journal on Marine Navigation and Safety of Sea Transportation*, 10, no. 2. DOI: 10.12716/1001.10.02.11. 287–299.

Wright, R. Glenn and Michael Baldauf. 2016c. Hydrographic Survey in Remote Regions: Using Vessels of Opportunity Equipped with 3-dimensional Forward-Looking Sonar. Journal of Marine Geodesy. 39, no. 6. DOI 10.1080/01490419. 2016.1245266. 439–357.

Wright, R. Glenn and Michael Baldauf. 2016d. Virtual Electronic Aids to Navigation for Remote and Ecologically Sensitive Regions. *The Journal of Navigation*. The Royal Institute of Navigation. doi:10.1017/S0373463316000527.

Wright, R. Glenn and Ian Russell. 2017. Navigation Sonar use in Maritime Frontier Exploration. *Soundings*. Hydrographic Society of the UK, Summer 2017, pp. 33–6.

Zaragoza, S. 2013. Spoofing a Superyacht at Sea, 30 July 2013. http://www.utexas.edu/know/2013/07/30/spoofing-a-superyacht-at-sea/.

# Chapter 7

# Communications

Communication to, from and within Maritime Autonomous Surface Ships (MASS) takes place in several forms and ranges in purpose from the internal exchange of navigation and engineering data and information, to bridge-to-bridge notification of intentions and the exchange of telemetry, imagery and operational data between ship and other locations necessary for management and control from a remote control center (RCC) or another ship. This is accomplished through a combination of visual methods, conventional radio, cellular and microwave links as well as satellite-based communications. MASS also creates new requirements for high bandwidth communication in the vicinity of the vessel. The complexity involved and the range and scope of sensors and automated systems pose a burden upon both internal vessel infrastructure and external communications channels to exchange unprecedented amounts of data and information. The management of such large volumes of data will not be possible without the allocation of additional radio spectrum bandwidth as MASS implementations proceed toward realization.

A summary of existing communications for ships in general is provided in this chapter, along with a description of current initiatives such as e-Navigation to extend and improve communication capabilities. This establishes the setting for discussions regarding the communication requirements of MASS that have their own set of unique needs and issues. The role of MASS in routine and emergency situations is discussed as well as the systems and the nature of the data they create, the channels through which data are communicated and the means by which this is accomplished to fulfill their goals and functionality needed in the roles and missions they perform.

## 7.1 GENERAL COMMUNICATION REQUIREMENTS FOR SHIPS

Since the invention of radio by Marconi in 1895, its potential use for communication on behalf of ships soon became evident, especially in distress situations as a result of two messages sent in 1899 to England over the span of two months. The first occurred on 17 March 1899 transmitted from the South Foreland Lighthouse when the merchant vessel *Elbe* had run aground onto Goodwin Sands off the coast of Kent [Telegraph 2017]. The second occurred on 30 April from the East Goodwin Sands Lightship on her own account when rammed by the steamship *R F Matthews*. Many additional accounts of passenger rescue occurred early in the 20th century from ships bearing the names of *Republic*, *Slavonia*, *City of Racine*, *Ohio*, *Minnehaha* and *Titanic* that resulted from radio distress calls.

In response to the *Titanic* disaster the first International Convention for the Safety of Life at Sea (SOLAS) was convened on 12 November 1913, and over several iterations was eventually to result in the latest version of 1974 with subsequent amendments [SOLAS 1974]. SOLAS Chapter IV specifies the requirements associated with radio communications for modern ships to establish distress, urgency and safety messages, as well as radio watches, bridge-to-bridge communications and general radio communications for operational and public correspondence traffic. The technologies that have evolved as a result of SOLAS include communications by medium frequency (MF), high frequency (HF) and very high frequency (VHF) transmitters and receivers, international navigational telex (NAVTEX) services for the delivery of navigational and meteorological warnings and forecasts and urgent maritime safety information (MSI), Inmarsat and Cospas-Sarsat satellite communications services, Global Maritime Distress Safety System (GMDSS), Automatic Identification System (AIS), Long Range Identification and Tracking (LRIT), Ship Security Alert System (SSAS) and other capabilities that are presently available.

MASS pose a unique challenge regarding safety of life at sea different than previous ship innovations in that these vessels are essentially unmanned. In and of themselves they pose little or no risk to their crews who operate them since they reside at remote locations such as an RCC. However, unmanned and automated ships can still wreak havoc on other vessels and installations at sea and on shore, and accidents in which they are involved may result in great economic and environmental damage. The communication requirements for MASS are greatly increased beyond the needs of conventional vessels to exchange the large volumes of data and information necessary to perform command and control functions from remote locations. The International Maritime Organization (IMO) e-Navigation initiative is presently considering new technologies to expand radio communications services for ships in general. However, although beneficial to MASS, these proposed new services are woefully inadequate to fulfill MASS communication needs.

## 7.2 e-NAVIGATION ENHANCEMENTS TO COMMUNICATIONS

The IMO e-Navigation initiative provides for the harmonized collection, integration, exchange, presentation and analysis of marine information on board and ashore by electronic means to enhance berth-to-berth navigation and related services for safety and security at sea and protection of the marine environment [MSC 85/26]. The implementation of e-Navigation provides for an effective and robust means of communications for ship and shore users of automated and standardized reporting functions for ship and voyage information transmitted ashore, sent from shore to shipborne users and information pertaining to security and environmental protection [MSC 85/26, p. 6].

Five e-Navigation solutions (numbered 1, 2, 3, 4 and 9) are prioritized in the Strategy Implementation Plan, each of which may be applied directly to the design of remote control center (RCC) facilities and the efficient exchange of marine data and information between ship and shore to all appropriate users [NCSR 1/28]. RCCs can benefit from the results of solution 1, which includes improvements in human factors and the human interface related to ergonomically improved and harmonized bridge and workstation layout, and solution 4, which focuses on the integration and presentation of available information in graphical displays received from communications equipment. Both MASS and RCCs can benefit from solution 2, which considers the means for standardized and automated reporting, solution 3, relating to improved quality, resilience and integrity of bridge equipment and, solution 9 for improved communications of Vessel Traffic Service Portfolios to identify possible communications methods most useful in different areas of operation (e.g., deep sea, coastal and port).

## 7.3 LIMITATIONS OF E-NAVIGATION AS RELATE TO MASS

Advances made through the course of e-Navigation over the past several years focus primarily on achieving incremental improvements and efficiencies in technologies conceived, created and developed during the last century including Electronic Chart Display Information System (ECDIS), Global Navigation Satellite System (GNSS), GMDSS, AIS, LRIT, and other navigation and communication systems. This includes the digitalization of many existing radio technologies and the introduction of new tools and methods for accomplishing the existing services provided and expansion of new services. These systems are and will continue to be critical to the advancement of both shipping in general as well as MASS. However, bandwidth requirements to support these existing services are already stretched to their limits and additional bandwidth is needed for further expansion to include the modification of existing and addition of HF and VHF channels plus new satellite services to support GMDSS [Bogens 2017; ISWG/ MASS 1/2/15].

The requirement for vastly expanded bandwidth to accommodate the communications needs of MASS will be fulfilled through the use of new methods and technologies that are rapidly becoming available in the commercial marketplace. Specifically, this includes new capabilities for cellular, satellite and microwave 5G ultrawide bandwidth services that will also accommodate Internet of Things (IoT) connectivity and cloud processing of data and simulation necessary to enable the development of many tools and methods needed to accomplish MASS. Details regarding the implementation of communications for MASS are provided in the paragraphs that follow.

## 7.4 COMMUNICATIONS REQUIREMENTS FOR MASS

Requirements for both internal and external vessel communications of MASS must include architectures capable of supporting different data formatting and transmission protocols that exist across engineering, navigation and other disciplines for various systems and sensors. This is necessary to support the ever-growing requirements for connectivity to send and receive information from remote and shore-based controllers and operators and to integrate with other vessels, unmanned vehicles, robotic devices and satellite services essential to MASS operations. Section 7.5 reviews the systems and types of communications achieved within the vessel itself, while Section 7.6 discusses the external functionality performed by these systems between MASS, other ships and the RCC. Included are roles of MASS in routine and emergency situations, the systems and the nature of the data they create, the channels through which data are communicated and the means by which this is accomplished to fulfill MASS goals and functionality.

Common across all levels of communication is the need for greater bandwidth to accommodate MASS data transfer rate requirements both internal to the vessel and externally to multiple users and destinations. Considering the combined data- and information-generating capabilities provided by onboard sensors, navigation and engineering systems, plus the need to communicate these data along with command and control instructions to operate MASS, data transfer rates are expected to exceed 1 Gbps (gigabits per second) generating exabytes (millions of terabytes) of information that needs to be communicated and managed effectively and securely. Such technological challenges are presently not achievable but are nearing fulfillment through the continued development and deployment of 5G technology within a wide range of devices across terrestrial, maritime and satellite networks.

Such change is not automatic in the maritime industry and requires planning and adoption on the part of the developers of ship-based systems and ship builders, ship owners and operators, and the regulators of the maritime industry. This is illustrated in part by some of the changes required to SOLAS Chapter IV ("Radiocommunications") corresponding to MASS that involve the addition, amendment, clarification and/or deletion of definitions, terms and regulations that presently exist. Examples of such changes include adding new definitions associated with automated systems, concepts associated with remote control and operation, handling communications at conning position, repeater operations for MF, HF and VHF communications, provisions for emergency situations and participating in search and rescue operations, new capabilities that acknowledge advancements in telecommunications, security provisions and the keeping of logs. Also to be considered are certification equivalencies of remote operators to onboard seafarers, including the certification of artificial intelligence-based systems that perform equivalent functions both on MASS and at remote locations such as the RCC.

## 7.5 COMMUNICATION INTERNAL TO MASS

A wide range of data originating from the suite of navigation instruments and environmental sensors is acquired on a continuous basis while underway, along with thousands of individual data points that represent parametric measurements of system factors that translate into health and performance metrics related to the various ship components, systems and machinery. At the top level a small portion of these data coalesce through modern integrated bridge systems (IBS) to provide convenient access to navigation, propulsion and control information and monitoring of essential factors relevant to the voyage. At a deeper level the vast majority of data is acquired, stored, organized and disseminated both internal to the vessel for use in optimizing onboard processes and providing full autonomy from an onboard reasoning system, and externally to shore for remote control and monitoring by technical, business and governmental interests. This trend toward acquiring and monitoring innumerable data representing all aspects of vessel navigation and performance increases exponentially with the introduction of MASS. Likewise, the complexity of internally communicating these data within MASS also increases and new methods and system architectures are needed to accomplish this.

### 7.5.1 Navigation Systems

The many integrated navigation system (INS) components perform a constant exchange of information to apprise the navigation watch of all relevant conditions to ensure safe and smooth navigation of the ship including compliance with the Convention of the International Regulations for Preventing Collisions at Sea (COLREG). This includes keeping a continuous watch of the sea using sight, hearing and all available means to report objects and conditions that may represent aids to navigation such as buoys, lights, land masses, natural and manmade objects, and other vessels and floating and fixed objects that may be considered as hazards to navigation that may harm the vessel.

The components of INS for conventional ships include all navigation functions representing heading, position and speed with corresponding electronic navigation chart (ENC) information, combined with requisite sensors and data resources for target identification and environmental awareness in terms of depths, aids to navigation and navigational hazards. The basic instruments comprising these components include dual ECDIS consisting of a master and backup unit, dual Radar systems with Automatic Radar Plotting Aid (ARPA) capability, positioning fixing systems, magnetic compass and gyrocompass, and echosounder. Additional capabilities can include autopilot and inertial navigation system, forward-looking navigation Sonar and enhanced sensors such as night-vision equipment. Using currently available off the shelf equipment and the NMEA 2000 data bus, data exchange rates are limited to approximately 250k bits per second (kbps) or approximately one-tenth the bandwidth afforded by mobile 3G communications technology introduced at the beginning of this century. The use of modern Ethernet local area network (LAN) data exchange technology provides considerably greater bandwidth approximating 10–100 megabits per second (Mbps), thereby providing the means for integrating additional equipment into the suite of navigation instruments.

### 7.5.2 Engineering Systems

Automated engineering systems perform the control and operation of the engines, generators, power distribution, drive train, fuel and lubricating systems, steering gear, and the many other system components and machinery. This is accomplished through continuous live monitoring of many thousands of data points within these systems that reflect pressure, temperature, flow, viscosity, power, torque, emissions, vibration and other factors, often indicated as functions of voltage, current and resistance. Additional sensors are also distributed throughout the ship to monitor various processes and events. These systems are constructed using a combination of high bandwidth fiber optic and wireless networks to provide low latency, reliable system and sensor data communications resulting in multiple integrated distributed networks that support redundancy to eliminate single point of failure events. Large volumes of data are coordinated and exchanged between system controllers with minimal latency and delay where its effects can result in inefficient control over independent objects, or oscillation and other behaviors that may prove damaging or destructive to propulsion system components and control surfaces such as rudders and fins used to reduce or prevent undesirable ship motion.

### 7.5.3 Imaging Systems

Vast amounts of live imagery must be communicated to enable onboard reasoning systems to achieve full autonomy and provide remote controllers with sufficient context of navigation and engineering data in terms of situational awareness both internal and external to the vessel. This includes the integration and fusion of live, high-resolution 360-degree visual imagery surrounding the close-in perimeter of the vessel for vessel security, port maneuvering, berthing and unberthing, and the nearby area out to 5,000 meters for aid to navigation detection and recognition as well as visual detection of vessel traffic to perform collision avoidance, and to detect hazards to navigation and potential threats. Additional imagery is also acquired for onboard surveillance to assist in maintenance inspections to detect leakages and spills, smoke, movement where there should be no movement and vice versa, and to determine the overall condition of machinery, the bilge, cargo holds and closed spaces. Imagery is also needed to inspect cargo and lashings and to provide security awareness related to key MASS hull access and control points. Further imaging requirements exist for a wide range of sensors generating pixel domain representations of physical phenomena from supplemental video, low light and infrared imaging systems, Radar, Sonar and many similar systems, and imagery from remote unmanned platforms and vessel-based robotic systems that operate in conjunction with MASS. Such volumes of imagery require 5G internal communications network data transfer rates of 1 Gbps or greater.

The fusion of integrated navigation and engineering systems data from power distribution and other systems together forms the basis for integrated bridge systems that can improve overall bridge management efficiency by providing many of the significant details of ship's navigation and engineering in one place. However, this is predicated on the assumption that seafarers are on board to visualize and interpret the meaning and significance of the data within the context of ship operations. With onboard reasoning systems performing fully autonomous operations and crew members for MASS residing at a remote location, it is necessary to supplement navigation and engineering data with environmental information sufficient to enable human operators and artificial intelligence-based reasoning systems to place themselves within the vessel on a virtual bridge to gain adequate situational awareness for proper decision making. This requires the acquisition and proper display of large amounts of live imagery that is orders of magnitude greater than the combined present-day volumes of data and information transfer requirements for navigation and engineering systems.

### 7.5.4 Local Area Networks

The rapid communication of large data volumes on the order of Gbps generated from navigation, engineering and imaging sources internally to the vessel is essential for intelligent reasoning systems to achieve autonomous operations. Likewise these same data must be routed into data pipelines for live transmission to the RCC and possibly other ships to facilitate the remote control of MASS. This can be accomplished internal to the vessel using the preferred method of fiber optic cables to achieve the necessary high data throughput rates, minimize security risks and achieve high reliability by eliminating interference from electromagnetic and other sources.

In locations on MASS that are hard to achieve access, where no cabling exists and for temporarily locating sensors external to the vessel and internally between decks, the establishment of wireless connectivity to devices is appropriate. This may be accomplished using "Wi-Fi", which is a trademarked phrase representing radio communication technologies for wireless local area networks [IEEE 802.11]. These wireless networks operate in the frequency bands of 2.4GHz, 5GHz, 60GHz and so on where signals may be blocked by metal decks, closed access doors and other impediments. This may be overcome through the use of Wi-Fi network signal boosters and repeaters placed throughout the vessel.

## 7.6 BRIDGE-TO-BRIDGE COMMUNICATION

When the subject of communication between ships is mentioned the conversation inevitably turns to discussion of VHF radio, which is the single most significant technology for achieving bridge-to-bridge voice communication. Its active use during normal operations is indispensable to communicate the intentions of vessels to other vessels such as in passing situations, and supplementary features to establish direct contact and emergency communications make this a highly useful tool to enhance safety of navigation. However, additional methods for communicating between vessels are mandated by regulation and applied in daily use. These methods include the use of light signals, day shapes as well as sound signals. Seafarers have significant training and years of experience in using the various visual, audible and technical tools and in understanding the significance and limitations of these methods. Indeed, COLREG Parts C and D and Annexes I-IV are dedicated to lights and shapes, sound and light signals, and technical details regarding their implementation [COLREG 1972].

In a discussion of communications involving MASS, nothing can be taken for granted and assurance must be given that all such methods are well defined, unambiguous and consistent in their use and application. Whereas it becomes second nature for properly trained and qualified seafarers to detect, recognize and react appropriately to all mandated methods of communication, no such assumptions may be made for MASS. Consideration of all aspects of communication must be made: from the "eyes" of MASS in terms of high-resolution visual sensors, to the "ears" that consist of sensitive microphones, the "extra sense" that makes possible the acquisition of electronic VHF communications and the understanding of its content in all its forms, and to the "mind" of artificial intelligence reasoning systems onboard MASS that can form comprehensive awareness of content, context and rationale through an amalgamation of these methods. The following paragraphs discuss the unique challenges MASS must overcome for safe and effective operation.

### 7.6.1 Lights and Shapes

All vessels operating at night and during periods of restricted visibility must display lights that signify the type of propulsion (power, sail or oar driven) and are to indicate if they are of special vessel type (air cushion, wing-in-ground (WIG) craft, surfaced submarine or sea plane). Lights are also used to indicate vessel orientation and length (side and stern lights, mast-head), function (fishing, towing, minesweeping) and operating limitations (not under command, restricted in their ability to maneuver, constrained by draft, anchored, aground). During daylight hours special shapes and configurations of these shapes are used to indicate many of these same attributes and circumstances.

The quality of data acquisition by MASS for lights and shapes is dependent upon the availability of properly stabilized imaging sensors (cameras) of sufficient resolution, sensitivity and capability to perceive colors (red, green, white, yellow) and distinguish between them, light mode (steady state, flashing) and shapes (diamond, conical, sphere) and to discern their number and orientation within the useful range of one to five nautical miles, depending on the type of light and vessel size. Cameras must also be placed appropriately around MASS to ensure 360-degree coverage is maintained, with backup capabilities in the event of camera failure to avoid blind spots in the field of view.

The methods used to detect lights and shapes and determine their meaning include the use of machine learning and statistical processes, with increasing reliance upon deep learning artificial intelligence as large enough data sets can be acquired to perform unsupervised training. Images acquired from cameras are analyzed using spectral filters and are based upon correspondence with training imagery data sets. The significance of detected lights and shapes corresponding to the vessel is determined within the context of other sensor systems through fusion of Radar, AIS and other available resources. Assuming adequate cameras are used in terms of range and resolution and proper imaging is achieved, the ability of MASS to properly classify and interpret their meaning is dependent on the degree of success achieved in the training of these systems. Watchstander qualifications include demonstrated knowledge and ability to perform these tasks in accordance with Standards of Training, Certification and Watchkeeping for Seafarers (STCW) requirements [STCW 1978]. However, STCW applies only to humans and for these experimental systems presently there are no standards for their training nor certification of reliable and consistent performance. Further, such systems are subject to the same limitations as humans in terms of degraded visibility due to precipitation, fog, smoke, and other weather and environmental conditions.

### 7.6.2 Sound and Light Signals

In addition to lights and shapes, active communication between vessels is accomplished using sound and light signals, alone or in combination. Sound signals are accomplished using the ship's whistle, bell and/or gong, depending upon vessel size. The whistle produces combinations of "short blasts" of about one second's duration, and "long blasts" of from four to six second's duration to reflect specific meanings that pertain to maneuvering and overtaking, restricted visibility, and warning signals. In some areas, sound signals may also be supplemented with light signals. Different frequencies of sounds are produced depending on vessel size, with smaller vessels less than 75 meters in length emitting a fundamental tone and/or one or more higher frequencies between 250Hz and 700Hz, vessels 75 but less than 200 meters to in length emitting a fundamental tone between 130Hz and 350Hz and vessels over 200 meters emitting fundamental between 70Hz and 200Hz These frequencies may differ for inland rules of different IMO member states. The ringing of a bell may be accompanied with the sounding of a gong on larger vessels to indicate a vessel at anchor.

The acquisition of high-quality sound information by MASS is dependent upon the use of highly sensitive microphone arrays placed around the perimeter of the vessel to ensure 360-degree coverage. Beamforming and adaptive steering methods are used to determine the approximate direction of the sounds, while acoustic filtering can reduce or eliminate noise and reflections. The acquisition of light signals is accomplished using the same stabilized imaging cameras and networks used for detecting lights and shapes.

Methods similar to those used in detecting navigation lights and shapes are available to detect sounds and light signals, and to determine their meaning and significance. Limitations also exist in the training and certification of reliable and consistent performance for such systems.

### 7.6.3 Signals to Attract Attention and Distress Signals

Essential to MASS is the capability to spot and recognize vessels needing help and in distress, even if able to render only limited direct assistance due to lack of onboard seafarers. This includes the capability to detect sound and light signals that are used to attract attention, including the beam of a searchlight, as long as they cannot be mistaken for an aid to navigation, an authorized signal identified in the previous paragraphs and avoiding the use of high-intensity intermittent or revolving lights such as strobe lights according to the COLREG. There are also prescribed VHF and/or HF radio distress calls in addition to visual and sound signals to specifically indicate distress including the use of red star shells and flares, flames on vessel, gunshots fired at intervals of one minute, waving arms and other visible indicators. MASS automated reasoning systems can readily learn to distinguish between these routine and extraordinary light, sound and radio signals using the same sensor suites already described, and to notify proper authorities of in extremis situations. In addition, MASS are in the peculiar position of providing live, detailed imagery and information to authorities not possible with conventional vessels. This can provide incident commanders with firsthand knowledge of on-scene conditions, thereby aiding in decision making to resolve the situation.

### 7.6.4 VHF Radio

The workhorse for bridge-to-bridge communications is the VHF radio, which is used by the master or person designated by the master to pilot or direct the movement of the vessel and to maintain a listening watch on the proper frequency(s) for calling and distress monitoring. Voice communication is used to provide information exchange needed for coordinating passing and overtaking situations. Digital Select Calling (DSC) is integrated into VHF radios to transmit non-distress query messages directly to other vessels.

MASS while underway must be obligated to act as relay stations with minimal communications channel latency to remote controllers based on other vessels or at the RCC to respond to queries and requests from nearby vessels. Furthermore, while operating in fully autonomous mode the vessel must be capable of replying on the radio in response to other vessel encounters in exactly the same manner as a human operator. The technology to accomplish this task is already available in the virtual assistants such as "Siri" on Apple phones, the Amazon "Alexa" and other similar products that already reside in many homes. The smart technology behind these products can be readily adapted to respond exactly as would a human being for VHF radio telephone use and, when connected to the MASS internal communications infrastructure, would have access to all bridge navigation equipment and other resources to draw correlations between the person speaking on the radio and the vessel from which the radio transmissions originate.

### 7.6.5 Distress Radio Communications

Digital Select Calling (DSC) is integrated into MF, HF and VHF radios to automatically transmit or receive alerts for distress, urgent or safety communications in the line of sight under the SOLAS convention. This is a core part of the Global Maritime Distress Safety System (GMDSS) that also includes sending and receiving maritime safety information via navigational telex (NAVTEX) messages. MASS must be capable of handling all conceivable maritime distress situations by acting as relay to remote controllers as well as acting on its own behalf through automated reasoning systems trained for emergency situations. This includes activating an Emergency Position Indicating Radio Beacon (EPIRB) to provide position information should it encounter distress at sea as well as activating a Search and Rescue Transponder (SART) to send Radar signals to X-band Radars useful to determining vessel location.

## 7.7 COMMUNICATION BETWEEN SHIP AND SHORE

The data necessary to be communicated between the ship and remote shore locations is accomplished by one or a combination of methods, depending upon the distance of MASS from terrestrial networks while underway. Should the vessel be inland or near the coast the use of cellular or mobile networks are likely to provide the least expensive solution as compared to satellite networks for the same volume of data. Medium/high frequency radio and/or satellite communications would generally be required while at sea. Where MASS are located in close proximity to port and at other strategic locations, it may also be possible to use microwave wide area networks (WAN) for data transfer. These first three options are discussed in this paragraph, while the many additional uses of WAN are described in detail in the next paragraph.

### 7.7.1 Medium/High Frequency Radio

In cases where communication is required beyond the line of sight capability of VHF radio, the use of Medium Frequency (MF) and High Frequency (HF) radio can provide much longer distance, and even worldwide communications between a ship and the shore and other vessels. This is typically accomplished using single-side band (SSB) modulation techniques that are much more efficient than traditional amplitude modulation (AM) used in the earliest shipboard installations. Maritime dependence for radio communications has been diminished somewhat with the advent of satellite communications which can provide far better bandwidth. However, unmanned and autonomous vessels can still make great use of MF/HF radio especially when supplemented with email, teletype, NAVTEX and weather facsimile services that can be integrated into the overall communications architecture.

## 7.7.2 Cellular Communication

Commonplace use of cellular and mobile communications began toward the end of the last century and continues to expand by providing new services at increasingly higher data rates. The pace of this expansion is accelerating and will continue throughout the 2020s with the installation of half a million new towers and small cell sites to support 5G data rates throughout the United States alone. These networks will support communication of the immense volumes of data and information MASS will generate, with the expansion in coastal urban areas likely to coincide well with the locations where these vessels will be used.

It should be noted that 5G cellular communications is accomplished at relatively low power levels where individual small cell cites having effective ranges of only about 150 meters or 500 feet. This may limit the availability of cellular networks suitable for the transfer of MASS data to only while the vessel is in port or when passing through inland waterways in close proximity to the shoreline. Beyond this range there is availability of cellular signals capable of providing data exchange at the slower 4G LTE data rates.

## 7.7.3 Satellite Communication

For greater distances and in remote locations, MASS can utilize existing satellite services such as those provided by Inmarsat, Cospas-Sarsat and others for the exchange of data and information. There is widespread subscription to these services, with Inmarsat Fleet Xpress supporting over 10,000 ships by January 2019 at speeds of up to 330 Mbps [Inmarsat 2017].

The ever-expanding broadband connectivity via satellite links is part of a concerted effort by several satellite services companies to launch tens of thousands of small satellite networks into low earth orbit (LEO) as well as highly elliptical orbits that will effectively provide ubiquitous broadband coverage across the entire planet, including the polar regions. These companies include SpaceX, Amazon, Boeing, OneWeb, LeoSat and others, each of which will launch and use their own satellites. Data exchange using LEO satellites will be advantageous for MASS as these channels exhibit very low latency rates due to the short distances between satellites and the users. The cost for data transmission is expected to significantly decline once several companies begin to compete for customers, with potential savings on the order of billions of dollars as compared to today's satellite communication costs.

## 7.7.4 Microwave Communication

Wireless wide area network (WAN) technology may be used to establish broadband connectivity between ships at sea and the shore to a range of up to 20 miles. This is accomplished through the use of stabilized microwave platforms that can track vessel locations and can aim the antenna directly to the ship. There are also several other uses for this WAN configuration applicable to MASS that warrant discussion in Section 7.8, as follows.

## 7.8 MASS AREA COMMUNICATIONS

The establishment of broadband communications capability for MASS at ranges of up to 20 miles from the vessel can provide numerous advantages and benefits that neither cellular nor satellite services can match, and with greater security since all communications are limited to between the participants involved.

One new capability involves establishing connectivity with one or more remote and autonomous surface and aerial vehicles (RSV/ASV and RAV/AAV) deployed simultaneously from MASS to receive high-resolution imagery and telemetry for self-inspection of the vessel. UAVs could also be used to scout ahead of the vessel to assess traffic congestion and to help detect potential hazards to navigation that may exist along the planned route of transit. This would also provide a means for MASS to assist in search and rescue efforts anywhere along its route.

Another advantage of a WAN is to provide broadband ship-to-ship communications that would enable the exchange of navigation and engineering data and information well beyond the requirements of typical bridge-to-bridge communications based upon VHF radio, visual and/or sound technology. For example, imagery of objects or events found to be significant by an automated reasoning system on MASS could be exchanged with another ship, conventional or MASS, on a reciprocal course in the event knowledge of such things were potentially useful to enhance safety of navigation or other reason. This could include the automatic transmission of swath bathymetry and/or side scan Sonar imagery for locations where changes in bottom configuration due to shoaling, uncharted wreck or reef, or other reason were detected that differ from or are not included in official navigation charts. This could also include sending forward-looking navigation Sonar and video imagery and location of shipping containers and other objects adrift at or below the waterline that these vessels should be aware to avoid.

Another new capability would include the ability to provide a means for command and control to enable MASS to be operated from another ship within the coverage area, and to coordinate MASS and tug operations while underway. WAN technology can provide yet another channel useful for offloading very large amounts of data and information with lower priority than live operational data. This includes background environmental data useful for crowdsourcing and data mining subsequent to the completion of each voyage.

Previous attempts have been made to achieve 4G LTE WAN connectivity onboard ships [Stone 2016]. While basic system functionality was accomplished, problems were encountered in terms of the general lack of space to install the number of antennas, amount of cabling and necessary hardware to properly implement these systems. Proposed 5G solutions may help to solve these problems.

## REFERENCES

Bogens, Karlis. 2017. GMDSS Modernisation and e-Navigation: Spectrum Needs. ETSI Workshop "Future Evolution of Marine Communication", 7–8 November 2017, Sophia Antipolis, France. International Telecommunication Union.

COLREG 1972. Convention of the International Regulations for Preventing Collisions at Sea. International Maritime Organization. London, 1972, with amendments.

IEEE 802.11. The Working Group for WLAN Standards, IEEE 802.11TM Wireless Local Area Networks. www.ieee802.org/11/.

Inmarsat 2017. Connected at Sea: Inmarsat's New High-Speed Broadband Service Hits 10,000-Ship Milestone. *gCaptain 2017*, 4 May 2017. http://gcaptain.com/fleet-xpress-exceeds-10000-ship-milestone-first-anniversary/.

International Convention for the Safety of Life at Sea (SOLAS) 1974. International Maritime Organization. http://www.imo.org/en/About/Conventions/ListOfConventions/Pages/International-Convention-for-the-Safety-of-Life-at-Sea-(SOLAS),-1974.aspx.

International Convention on Standards of Training, Certification and Watchkeeping for Seafarers (STCW) 1978. As amended, including the 1995 and 2010 Manila Amendments.

ISWG/MASS 1/2/15. Summary of Results of the First Step of the RSE for SOLAS Chapter IV. General comments on degrees of autonomy three and four. Turkey. International Maritime Organization. 1 August 2019, Annex 1, 2.

MSC 85/26. Strategy for the Development and Implementation of e-Navigation. International Maritime Organization. London, MSC/85/26/Add-1, Annex 20.

NCSR 1/28. Draft e-Navigation Strategy Implementation Plan. International Maritime Organization. London, Annex 7.

Stone, Adam. 2016. Navy Rethinks How to Bring 4G LTE to Sea. C4ISRNET, 12 February 2016. https://www.c4isrnet.com/show-reporter/afcea-west/2016/02/12/navy-rethinks-how-to-bring-4g-lte-to-sea/.

Telegraph 2017. The First Ever Radio Distress Call is made off Kent Coast. *The Telegraph*. London, 17 March 2017. https://www.telegraph.co.uk/technology/connecting-britain/first-ever-radio-distress-call/.

# Chapter 8

# Security

The nature of Maritime Autonomous Surface Ships (MASS) being without onboard seafarers to personally oversee real-time events and, with the exchange of large volumes of data and information between the vessel and a remote control center (RCC) situated on shore or on another vessel, provides many opportunities where physical and cyber security vulnerabilities may be exploited. Potential threats may be encountered anywhere through the entire MASS ecosystem including at the vessel itself, the RCC and the communication channel connecting the two locations. This chapter discusses the potential threat vectors for MASS, their implications for security, and potential countermeasures that may be undertaken to minimize their risk and to overcome these threats.

## 8.1 IT BEGINS WITH THE VESSEL SECURITY PLAN

The shipping company designates a Company Security Officer (CSO) in accordance with nation state and International Maritime Organization (IMO) guidelines with several areas where minimal competencies are established. These include strategies, plans and tactics to:

- Develop, maintain and supervise the implementation of a Vessel Security Plan (VSP);
- Assess security risk, threat and vulnerability;
- Ensure that security equipment and systems are properly operated, tested and calibrated; and
- Encourage security awareness and vigilance. [MSC/Circ. 1154]

The International Ship and Port Facility Security (ISPS) Code requires the development of a VSP that prescribes protective measures for each of three security levels to be carried out by crewmembers [ISPS 1988]. These include Maritime Security (MARSEC) Levels 1 and 2 for routine and heightened security and Level 3 for a pending incident or when an incident has already occurred. Additional competencies are presented that reflect procedures required to be performed by the Vessel Security Officer (VSO), which are discussed in the next paragraph.

The general outline for the VSP cites procedures and security measures tailored to the vessels themselves, the routes that they operate and the ports to which they call. As applied to MASS, there is general agreement that the ISPS Code can be adapted to fulfill its requirements through the addition of appropriate definitions and the use of International Convention on Safety of Life At Sea (SOLAS) alternative and equivalent security arrangements [ISWG/MASS 1/2/13]. This includes the VSP that still applies regardless of vessel operating mode – conventional, remotely operated or autonomous. The essential elements of the VSP must continually be updated to reflect new threats as well as new technologies to mitigate these threats. This includes modern physical threats as well as the ever-expanding realm of cyber threats. However, the unique requirements and characteristics of MASS that must also be considered throughout this process are discussed in this chapter. A comprehensive and detailed list of criteria is available for evaluating all aspects of a VSP as implemented by the U.S. Coast Guard and other national authorities considering the Vessel Safety Assessment, vessel details; qualifications and responsibilities of the Master, CSO and VSO; drill and exercise requirements; and the many other aspects of vessel and facility security [USCG 2017].

Normally required to be kept onboard, in the absence of onboard seafarers it is presumed the VSP would be kept at the RCC. For fully autonomous operations the VSP equivalent would need to be codified in some manner through computer software such that an automated reasoning system would be able to access its contents for decision making while operating in fully autonomous mode in terms of interpreting sensor data and information for threat awareness and reaction to threats.

## 8.2 MASS ABILITY TO MAINTAIN SECURITY

On conventional ships there are three levels of seafarer security responsibilities and awareness cited in the International Convention on Standards of Training, Certification and Watchkeeping (STCW) including endorsements for VSO, Vessel Personnel with Designated Security Duties (VPDSD) and Security Awareness (SA) [STCW 1978a]. These same levels are also appropriate for MASS but, absent onboard seafarers, these duties must be performed during remote operations by qualified staff. During fully autonomous operations these responsibilities must also be passed along to an onboard automated reasoning system capable of performing the duties associated with these positions. Similar requirements to maintain physical and cyber security exist at shoreside installations such as the RCC. This paragraph looks at the specific qualifications needed to perform these duties as well as the capabilities that may be available to assist in their accomplishment.

### 8.2.1 Qualifications

The minimal competencies established for VSO, the highest level of onboard security authority, are essentially equivalent to the requirements for CSO under an already existing and approved VSP [ISPS 1988, Chapter XI-1/8; STCW 1978b]. This includes competencies needed to maintain and supervise VSP implementation; assess security risk, threat and vulnerability; ensure that security equipment and systems are properly operated, tested and calibrated; and encourage security awareness and vigilance. One additional requirement exists for the VSO to undertake regular inspections onboard the vessel to ensure security measures are properly implemented and maintained. Competency to perform inspections can be demonstrated through the development of automated reasoning systems capable of detecting anomalous conditions as may be presented by sensor data and imagery.

The methods used for demonstrating competence for human candidates for CSO and VSO positions include the assessment of evidence obtained from approved training or examination, and/or approved experience and examination including practical demonstration of competence. This is evaluated according to procedures and actions in accordance with ISPS Code, SOLAS, 1974 (as amended), and other legislative requirements. At present there are no established requirements for demonstrating the knowledge, understanding and proficiency for shipboard automated reasoning systems to evaluate their competence in performing the required security tasks other than to apply the same standards as for humans. This will prove difficult for the testing of knowledge that is greatly dependent upon the methods used to train these systems. Difficulty also exists due in part to the complexity of many machine learning methods along with the sheer impossibility of verifying deep-learning artificial intelligence architectures due to our lack of visibility into the elements that make up their architectures. This is a glaring technical deficiency where, at present, we cannot even know if the "knowledge" they use as represented by the connections between internal neurons of artificial neural network arrive at the same conclusions using different methods given the same criteria, or how different outcomes are substantiated by traceable rationales. In the future this will likely improve as new architectures providing greater insight into their decision-making processes are developed. At present, in the absence of an ability to gauge knowledge, these reasoning systems must be tested using "black box" methods based entirely upon the results achieved given specific situations without considering how they arrive at the outcome. This is a distinct and fundamental change from human capability assessment that is based in part upon the testing of general and specific knowledge of security-related topics and policies.

## 8.2.2 Capabilities

There are many facilities available onboard ships enabling the VSO, human or automated, to demonstrate initial competence and carry out ongoing drills in the actual performance of safety and preventive measures. The capability to perform inspections can be demonstrated through the use of fixed and robotic sensors that provide data and imagery collected throughout the vessel communicated to one or more automated reasoning systems designed to analyze and interpret this information. This is accomplished in terms of weapons, devices, methods and people intent on compromising vessel safety and security. Such facilities relate to physical security and structural integrity, threats both internal and external to the vessel, procedures, radio and telecommunications, cyber security and other areas that may pose a risk to people, property or operations onboard MASS or within the RCC. Detailed discussion on these and related topics appears in the paragraphs that follow.

## 8.3 PHYSICAL SECURITY

The purpose of physical security is to protect the safety of personnel, physical assets and facilities of a company or organization. For MASS this includes to the vessel itself and extends to the RCC, to the routes and ports within which MASS operate. Since the physical security of land-based personnel and facilities are adequately covered in other texts, this paragraph focuses on the vessels themselves represented as the object of the VSP. Topics addressed include the features and systems that may help in preventing the entry into MASS spaces by unauthorized personnel, taking over of vessels through acts of piracy and reducing the exposure of MASS as a viable target for attack.

The security requirements for MASS differ with each implementation based upon the type of vessel, its mission and purpose, and the environment in which it operates. The following paragraphs list some of the possible security precautions that may be taken to secure the vessel itself, some of which are also applicable to the RCC. Specific threats and security requirements should be identified in the VSP.

### 8.3.1 Unauthorized Entry

The prevention of unauthorized entry onto the vessel itself as well as to the internal spaces is the first step that must be taken to properly secure the vessel. This is accomplished using four different levels of defense based in part upon the stage at which the detection of potential threats and unauthorized personnel is accomplished. The first level is where one or more automated reasoning systems using smart sensors with integrated defense mechanisms onboard MASS would be capable of detecting potential threats in the general area surrounding the vessel before a boarding attempt is attempted. Failing that, the second level is the immediate perimeter adjacent to the vessel where boarding is attempted and onboard defenses may prevent a successful boarding from being accomplished. If one or more unauthorized persons come alongside and manages to board, the third level of defense should thwart all attempts to breach the external defenses and enter the interior spaces of the vessel. Lastly, once onboard and exterior defenses are breached allowing for entry into the interior of the vessel, these spaces should include multiple internal defensive systems and the design of MASS should be sufficiently unique and confusing to those unfamiliar with the vessel such that essential engineering, navigation and other systems distributed around the vessel are difficult to find and impossible to breach.

### *8.3.1.1 Area Defenses*

The capability to detect potential threats in advance of their becoming actual threats is dependent upon the type, resolution and timeliness of sensor inputs and available sensor data. However, this is inextricably linked to the amount of available knowledge about potential threats and the training that is available to an operator at a RCC or an onboard reasoning system regarding threat awareness. It is also greatly dependent upon the density of vessel traffic in the area in which the vessel operates.

Area defenses must begin with gaining awareness through the integration of all available long-range sensors such as Automatic Identification System (AIS), Radar, Sonar, visual and real-time satellite resources to create a comprehensive real-time picture of the threat environment considering not only the characteristics of nearby vessels but also their behavior in relation to MASS while underway. At present it is possible for an operator to create an overall threat assessment for a general area by viewing nearby vessels within a radius of approximately three to ten nautical miles and obtaining relevant information to ascertain their size, type, heading, speed, closest point and time of approach and other characteristics. Their overall threat potential and capacity to become a threat over the span of the next 3–30 minutes is dependent upon their speed, whether they are moving toward or away, and if abrupt changes in course and speed take place to increase the threat level. This includes whether the type of vessel can be characterized as being historically involved in pirate or other hostile activities such as a mother ship and the high speed and highly maneuverable small craft and personal watercraft they can harbor that can come quickly alongside and attempt to board or place explosive charges on the exterior of the hull.

The required knowledge and the task of analyzing high levels of operational detail in areas where there are more than a few vessels make continuous efforts for area threat assessment an overwhelming task for an individual person. Automation of threat assessment is the key to ensuring the viability of MASS, especially where one operator is responsible for simultaneously controlling and coordinating the movements of several vessels. For future conventional and autonomous ships further research is needed to develop a set of tools to build datasets of specific known hostile vessels, general vessel types and vessel behavior models to train and enable automated reasoning systems to assess and rank threat levels associated with individual vessels. Such tools can prove to be invaluable to remote operators. These tools would also be indispensable for automated reasoning systems onboard MASS for navigation and operation to assist in its own defense should the need arise, and to prepare to react to potential threats by developing a real-time plan to implement all appropriate defenses within a predefined perimeter. The size of this perimeter is likely to be dynamic as a function of how many vessels are in the immediate area and its geographic constraints.

In most cases, except when in the open ocean or in areas where there are few vessels, it would generally be ineffective to consider taking active defensive measures against vessels within this three- to ten-mile zone as many false positives would occur. Should this perimeter be penetrated, communication of appropriate data and imagery must be made to the RCC and all defensive mechanisms must be moved from a standby condition and placed into an operational mode ready for immediate deployment and use. An exception would be for small high speed craft that can rapidly close short distances, which would be considered a higher threat than most other vessels. However, the judicious use of extra speed and a precautionary change in course to increase space between vessels in the event of suspicions behavior can provide an initial advantage to detect and overcome potential threats before they become actual threats. If vessels that are potential threats also make corresponding changes, such actions would significantly raise the threat level.

### 8.3.1.2 Perimeter Defenses

As a potential threat closes to within a perimeter of three miles its threat potential dramatically increases since it is now approaching the range of many offensive weapons that may be used to threaten MASS, and preparations to repel boarders must now be made. The threat may take the form of a single vessel or a swarm of vessels that may attempt to overwhelm defensive capabilities that may be used against them. Smart ship sensors and sensor data provided to the remote operator and/or onboard reasoning system has already resulted in the identification and prioritization of potential threats within the perimeter whereby automated defensive measures may be initiated while they are still at a distance and not yet alongside. Communication of the increased threat level should also be made to the RCC to specifically advise appropriate personnel of the increase in threat status.

MASS in general should be somewhat immune from the usual focal point of attack being the bridge since it is unmanned. However, defensive measures shown to be effective may now be activated including the use of non-lethal directed-energy weapons that can disorient and temporarily blind and deafen personnel onboard threatening vessels. One such approach using a laser weapon was demonstrated as early as 2011 by BAE Systems and is increasingly being deployed on the naval vessels of several countries to help deter piracy [Gray 2011]. A second approach is the use of high power acoustic systems such as the Long Range Acoustic Device of the American Technology Corporation that can communicate directly to other vessels as a very loud hailing device, and can also direct extremely loud sounds at up to 150 decibels that are capable of disorienting and temporarily deafening personnel on threatening vessels [Evers 2005]. Another invention resulting from naval experiments is a non-lethal directed microwave weapon known as Mob Excess Deterrent Using Silent Audio (MEDUSA) that creates excruciating sound inside the head [Hambling 2008]. A variation of this technology includes the Silent Guardian system, developed by Raytheon, that uses directed energy to heat target surfaces, including human skin, causing pain with a range up to 250 meters [Hambling 2006]. A much lower technology yet effective approach for close in encounters is the use of water cannons to discourage any attempts to get directly alongside in an attempt to board the vessel.

All weapons such as these are capable of being controlled remotely from a RCC and onboard under autonomous control. The devices listed above are not new, and refinements to these technologies have taken place over the years to where they can provide effective means of defense. However, should an intruder manage to overcome one or a combination of these perimeter defensive measures and is able to come directly alongside, the next set of defensive weapons designed to thwart breach of the exterior of the vessel may be brought to bear.

### 8.3.1.3 Exterior Defenses

Precautions that may be taken to deny access to deck and vessel spaces pertain to all phases of MASS utilization and include during the voyage itself, while in port and when laid up for repair. This applies to the hull exterior as well as all doors and hatches accessible to interior compartments that may be manned by other than seafarers as well as machinery and cargo spaces. Where a threatening vessel has come along side and boarding is being attempted some of the perimeter defenses may be ineffective due to the close proximity of the target and threat vessels. The use of remotely and autonomously controlled water cannons and fire hoses may still be effective in deterring boarding as may the use of automatically deploying nets that can foul small boat propellers and prevent their operation. Additional products such as compressed air guns that fire rubber bullets and other projectiles and the use of slippery foams and foul-smelling liquids are also available to deter boarders. Should the boarders manage to avoid or break through these defenses and are able to scale the hull, the last line of defense can include variations of electric, barbed and razor wire and other features unique to the design of MASS that can deter gaining access to the deck.

Once access to the vessel exterior and deck has been attained there should be few portals through which routine entry may be made, with all other portals secured through automatic locking mechanisms that cannot be opened external to the vessel by other than properly credentialed authorized personnel or a remote operator at a RCC.

Routine security measures such as physical and electronic checks of personal credentials, facial recognition, fingerprint and/or iris scans and other methods may be performed while underway, in ports and during maintenance to deny access to unauthorized personnel. Notwithstanding malicious behavior on the part of insiders, the assumption behind these security methods is that anyone passing these checks will be known to the shipping company and cleared for access. Likewise, unauthorized personnel will not pass these security checks and exterior portals should remain secure. Throughout this period communications can be maintained with the RCC providing live imaging to provide real-time situational awareness that may help to identify known threats and aid in their eventual capture.

### *8.3.1.4 Internal Defenses*

Once unauthorized personnel gain access to the deck and the interior of the vessel has been entered, these spaces should include multiple internal defensive systems such as physical constraints that prevent further access, and strobe lights, tear gas, acoustic and other measures that can make the experience highly unpleasant, disorienting and even painful. The design of MASS should also be sufficiently unique and confusing such that essential engineering, navigation and other systems distributed around the vessel are difficult to find. Even if found there should be physical impediments that are impossible to breach and would prevent connecting hardware devices or installing software that would make it possible to take control of the vessel. The ability to seize the vessel should also be drastically reduced by the lack of qualified onboard seafarers to threaten unless they comply with the demands of attackers. This may not preclude taking the vessel under tow. However, the logistics of such operations tend to be beyond the capabilities of most terrorists and pirates without nation state sponsorship.

### *8.3.1.5 Non-lethal Defenses*

The above concepts all entail the use of non-lethal defensive measures that tend to avoid most of the controversies associated with the use of onboard armed security personnel. Most importantly, the risk that the use of automated defensive measures may go out of control and have unintended consequences for innocent bystanders is also minimized by this approach.

Indeed, the entire concept of lethal force against human targets being ordered by an automaton is beyond the scope of this chapter and best left to others to work out such details.

### 8.3.2 Physical Attack

Other than the defensive measures described in the previous paragraph it is unlikely that MASS of any design or function will be able to withstand a direct attack with intent to disable or sink the vessel. The onboard sensor systems should be very adept at detecting that an attack may be imminent or is underway, with corresponding notification through multiple modes of communication to alert the RCC and the proper authorities in a timely fashion. This includes providing real-time data and imagery obtained using visual sensors to detect incoming threats above the waterline such as rocket-propelled grenades, as well as Sonar sensors capable of detecting and characterizing scuba divers, mines and other surface and underwater threats and providing data useful for forensic investigation.

## 8.4 THREATS INTERNAL TO THE VESSEL

There exist threats that pose as trusted personnel which, by their presumed trustworthy status, can overcome many of the defenses described thus far in this chapter. These may include company employees and vendor personnel performing essential duties onboard the vessel as well as at the RCC. In the process they may be able to sabotage defensive mechanisms that will promote compromise such as giving away codes and passwords to security systems that allow future access, and by introducing spyware and malware into computer systems that place data and information at risk and may even facilitate the remote hijacking of a vessel.

### 8.4.1 Proper Vetting of Authorized Personnel

The vetting of company and vendor personnel for MASS does not differ appreciably from the needs for standard shipping and industrial practices to establish initial and continuous trust that can identify risks regarding a trusted insider. However, the means by which this is established are likely to change dramatically in upcoming years to coincide with changes in technology. This includes routine survey of social media sites such as Facebook and Twitter as well as industry blogs and news sites looking for facts, opinions and comments from and about individuals undergoing vetting. This information in and of itself may or may not be considered threatening but nonetheless may raise flags regarding potential trustworthiness and internal threat issues. In the vetting process it is very likely that passwords to personal social media accounts will be requested by employers to review account content history over several years as part of their vetting process.

Official vetting through nation state and local port authorities must also be accomplished. For example in the United States one measure taken to perform the vetting task is establishment of the Transportation Worker Identification Card (TWIC©) required under the Maritime Transportation Security Act for workers who need access to secure areas within maritime facilities and vessels [TSA 2019]. This provides a means to conduct a security threat assessment and background check to determine eligibility. In the case of MASS the term maritime facilities not only applies to the ports in which MASS operate as well as the vessels themselves, but also for the RCC and other remote locations from which MASS and its systems may be accessed, controlled and operated.

Additional criteria for the vetting of authorized of personnel are established through parts A and B of the ISPS Code, the Guide to Maritime Security and the ISPS Code, and Guidance for the Development of National Maritime Security Legislation [ISPS 2012; MSC.1/Circ. 1525].

### 8.4.2 Obsolete Software

MASS by its very nature is software driven and highly dependent upon the programming that controls the many onboard and RCC systems. Security associated with software and computer systems is covered later in this chapter under the topic of cyber security where automated processes and manual procedures are established to minimize risk related to data theft and the introduction of malicious software or malware. However, special attention is given here regarding the potential for trusted personnel to install software that has been authorized for use, yet may be obsolete or contain differences and/or modifications that have not been demonstrated to be interoperable with other system software products. This can include other software configurations using the same network(s) that may cause operational errors and degraded performance, software to eliminate known problems, or software is beyond their end of life and is no longer supported or receiving security updates.

For example, the Microsoft Windows 7 lifecycle ended in January 2020. Yet as of the beginning of 2019, 43% of businesses still ran Windows 7 despite the better security, efficiency and features associated with later versions of this software [HelpNet 2019]. Even worse, systems running Windows XP still exist despite being long past the end of its lifecycle [Hruska 2019]. Yet I have personally witnessed the use of XP and NT on naval vessels as late as 2016, with some commercial vessels still using this version.

A prohibition on the use of obsolete software is essential and must be established and enforced. This is especially necessary for MASS applications where sensor intensive systems are controlled and/or monitored by shore-based operators or onboard reasoning systems which must then process and communicate large amounts of data and imagery in real time. Such a statement seems on its face to be obvious and the need without question. However, the testing needed to ensure software interoperability and robustness is not easily achieved and takes a great deal of time, effort and funds particularly in specialized applications such as MASS that do not have massive consumer use where tens of thousands of installations are examined by many sets of eyes. Attempts to properly test software are often met with resistance as new, more capable software products and upgrades are released and pressures to adopt later software configurations are hard to resist. This further expands the problem and results in a never ending cycle where the actual properties and performance of off-the-shelf software continue to change with unknown effects.

### 8.4.3 Crew and Vendor Awareness

The very first paragraph in this chapter emphasizes the need for proper security training and awareness on the part of the crew and other staff members who operate, maintain and repair MASS as specified in the Vessel Security Plan. However, the VSP covers the shipping company itself and possibly not all organizations and third-party vendors with which it must interact and do business. Of equal importance is the level of security awareness on the part of each of these vendors as they supply vessel system hardware spanning all MASS operations including engine and propulsion, navigation and communications, and the software that drives these systems. One source cites figures stating that 60% of all data breaches that occurred in 2015 were caused by third-party vendors that potentially affected more than 41 million people [Belding 2019].

It is essential for shipping companies involved with MASS to ensure all vendor and personnel from other organizations document their employee training practices as being adequate for security awareness. This includes the establishment of Service-Level Agreements that specify security relationships between the parties that are comprehensive and enforceable. Absent such provisions, there will be glaring deficiencies in facility and vessel security plans that can easily be overcome and provide a welcome mat across the back door entrance to MASS facilities and vessels.

## 8.5 EXTERNAL ELECTRONIC THREATS

There exist external threats to MASS of an electronic nature that can easily lure any such vessel off course and into the territorial waters of a hostile state to be captured, to be hijacked without having to approach or board the vessel, or to cause a vessel to go aground or otherwise harm itself and/or other vessels and waterfront installations. Another threat is the cloaking of potentially hostile vessels with legitimate identities to help catch a vessel unaware. These are truly matters of security which are discussed here as a brief reminder even though these threats were dealt with in Chapter 6 on Navigation.

### 8.5.1 GNSS Spoofing and Denial of Service

The ability to spoof Global Navigation Satellite System (GNSS) receivers into showing geographic location coordinates that are inaccurate and misleading is well documented and discussed in Section 6.3.6.1 of Chapter 6. Hostile entities can use this to their advantage to hijack a vessel and send it to a nearby location of their choice, which may be a reef or shoal, without ever stepping onboard merely by broadcasting interference signals from another vessel or close by land location. This can be accomplished gradually so that smooth and small transitions indicated on Electronic Chart Display Information System (ECDIS) are likely to escape notice by crewmembers which may be onboard or at a RCC. At present there are no automated methods to detect such a local event, yet symptoms exist within sensor data that can make these events detectable including differences in bottom depth contours between the perceived location and the spoofed location from the perspective of actual depths, changes in depth and the rate of change in depth [Wright 2019; Wright and Baldauf 2016]. The development of automated tools to detect such spoofing events that can be integrated into ECDIS can provide great assistance in reducing this risk. The same techniques may be applied as a backup method of navigation in the event of denial of service attack that renders GNSS inoperable.

### 8.5.2 AIS Limitations

The ability to spoof, jam and overcome Automatic Identification System (AIS) capabilities due to clutter are significant limitations of this technology. Relying on it to determine the presence and location of nearby vessels may be detrimental to the security of a vessel. Discussed in Section 6.3.6.2 of Chapter 6, these limitations may be exploited to make a vessel appear to have different characteristics, e.g., length, size, name and home port, and to show its presence at a location that may be miles away from its actual location.

Vessels that may be potential threats are unlikely to use AIS, and alert visual watchstanding procedures must be combined with electronic watches such as Radar to ensure potential threats are detected. However, if the use of AIS comprises a significant part of a VSP, then awareness of its potential limitations are crucial to its usefulness and effectiveness as a tool.

## 8.6 CYBER SECURITY

Commercial vessels of all types are increasingly the targets for infiltrating shipboard computer systems in attempts to gather data and introduce malware that can reduce their functionality and possibly disrupt critical onboard systems. While most attacks have not directly affected essential control systems of conventional ships, MASS are much more reliant upon computer systems and networks for their operation and therefore the consequences of successful cyber attacks can be devastating. The problem also exists at remote facilities such as a RCC that will become prime targets along with the terrestrial and satellite networks that support MASS and RCC communications.

### 8.6.1 Incidents

Cyber attacks on worldwide infrastructure and facilities can be numbered in the tens of thousands on a daily basis being carried out by criminals, adversaries and nation states. When successful, a data breach may take up to half a year to even detect that an attack has occurred [Galov 2019]. With the ever-expanding Internet of Things (IoT), attacks on mobile and smart devices such as cameras and smart sensors connected via the Internet are also rapidly increasing – up over 600% in 2017 from prior years.

With the digitalization of the shipping industry cyber attacks are becoming much more prevalent. A breach of security that resulted in a February 2019 incident onboard a deep draft vessel on an international voyage to the Port of New York and New Jersey severely impacted their shipboard network exposing critical vessel control systems to significant vulnerabilities [USCG 06-19]. With many crew members using their shipboard Internet connections to perform communication by email and to view website content, phishing attempts and the download of malware targeting commercial vessels is also increasing [USCG 04-19]. The cyber security firm FireEye in 2018 reported tracking nation state sponsored intrusions spanning many years targeting mostly United States engineering and maritime entities using enhanced malware toolsets capable of broadly accessing systems through backdoor vulnerabilities, cracking passwords and codes, infiltrating and compromising web sites, and exfiltrating large volumes of data [FireEye 2018].

One of the boldest and costliest cyber attacks perpetrated on the maritime industry affected A.P. Møller-Maersk beginning on 27 June 2017, shutting down operations that required a complete rebuild over the next ten days of their entire global network infrastructure consisting of 4,000 servers and 45,000 computers [Greenberg 2018]. The cost of this incident was estimated to be between $250 and $300 million dollars. However, A.P. Møller-Maresk was only a small part of a massive cyber attack that originated from a small software firm in the Ukraine and spread worldwide affecting millions of computers at some of the world's largest companies with a total cost estimated at around $10 billion dollars. While all of Maresk's ships were affected by the shutdown as a result of company and port computer system unavailability, none of the computer systems on these ships were infected with the NotPetya malware found to be the cause of the incident. This incident demonstrates the sheer magnitude of destruction that can take place as a result of global interconnectivity over broad business lines where cyber security is very complex. This incident probably also demonstrated the benefits of ships that are not connected to the Internet.

### 8.6.2 Implications for MASS

The implementation of MASS will be unprecedented in terms of transportation system complexity considering the number and types of interconnected computer systems, automated systems providing shipboard control, smart sensors, Internet of Things appliances, and the terrestrial and satellite networks through which communication are established between the vessel and operators at a RCC. Each one of these elements provides multiple nodes through which cyber attacks can be launched. Judging by the Maresk incident, the potential for monetary loss for poorly protected computer networks and systems is huge. While there was no loss of life that could be directly attributed to the Maresk attack, a future attack on a computer network that controls seagoing vessels can set the stage for massive loss of life and environmental disaster associated with the vessels themselves as well as other vessels, ports and everywhere along the routes where they operate.

### 8.6.3 Cyber Security Program Implementation

The topic of maritime cyber security has become very important in recent years and much guidance exists for implementing effective programs. Guides have been produced and supported by BIMCO, American Bureau of Shipping (ABS), the U.S. National Bureau of Standards (NIST) and other groups for the implementation of maritime cyber security practices that provide useful information for establishing such programs [ABS 2018; BIMCO 2018; NIST 2016]. These guides address the significant issues for securing the safety of critical infrastructure through the development of frameworks and policies.

The IMO has also developed guidelines that present high-level recommendations on maritime cyber risk management to safeguard shipping from current and emerging cyber threats and vulnerabilities [MSC-FAL.1/Circ. 3]. This identifies the functional elements that support effective cyber risk management through concurrent and continuous practices that relate to the identification of personnel roles and responsibilities and systems, assets, data and capabilities that pose risks to ship operations if disrupted.

This also addresses the protection of shipping operations through the implementation of risk control processes, measures and contingency planning; timely detection of cyber events, responding to restore systems and recovery measures.

The U.S. Coast Guard encourages the performance of cyber security assessments to better understand the extent of cyber vulnerabilities and has provided recommendations for vessel and facility owners, operators and other responsible parties to improve cyber security [USCG 06-19]. Recommended measures include:

1. Network segmentation in sub-networks that make it harder for an adversary to gain access to critical systems and equipment;
2. Elimination of generic login credentials that can be used by multiple individuals;
3. Highly restrict the use of external media, and especially USB drives, that can host a wide range of malware;
4. Installation of antivirus software; and
5. Regular patching of software to incorporate updated software into computer systems, and especially security fixes to old vulnerabilities.

The last segment of the vulnerability chain is the terrestrial and satellite communications channels where lack of security can expose ships to remote hacker attacks that can allow for their takeover by permitting access to ECDIS and autopilot functions [Kovacs 2018]. This is done by exploiting vulnerabilities such as the use of default credentials for satellite terminals to allow unauthorized users to gain administrative-level access. The use of strong administrative passwords would help solve this particular problem. However, the satellite terminal can facilitate access to the ECDIS and engineering systems. Many security problems were found across over 20 different ECDIS systems that included using obsolete software such as Windows NT. Similar problems were also detected in engineering systems used to control steering, engines, generators and other ship components that relate to communications using the NMEA 0183 bus that does not use any authentication, encryption or validation protocols. This allows attack through modification of bus data that would enable course changes when the autopilot is engaged.

## REFERENCES

ABS 2018. Cybersecurity Implementation for the Marine and Offshore Industries, ABS Cybersafety™. American Bureau of Shipping, Houston, Texas, 15 June 2018.

Belding, Greg. 2019. Security Awareness for Vendors and Contractors. Security Awareness, 4 February 2019. https://resources.infosecinstitute.com/security-awareness-for-vendors-and-contractors/#gref.

BIMCO 2018. The Guidelines on Cyber Security Onboard Ships. Version 3. www.bimco.org/news/priority-news/20181207-indistry-publishes-improved-cyber-security-guidelines.
Evers, Marco. 2005. Sonic Canon Gives Pirates an Earful. *Spiegel Online*, 15 November 2005. https://www.spiegel.de/international/spiegel/the-weapon-of-sound-sonic-canon-gives-pirates-an-earful-a-385048.html.
Galov, Nick. 2019. Cyber Security Statistics for 2019. *Cyber Defense Magazine*, 21 March 2019. Source: ZD Net as reported by Galov. www.cyberdefensemagazine.com/cyber-security-statistics-for-2019.
Gray, Richard. 2011. Laser Cannons to Defend Ships from Pirates. *The Telegraph*, London. 9 January 2011. https://www.telegraph.co.uk/news/worldnews/piracy/8247665/Laser-cannons-to-defend-ships-from-pirates.html.
Greenberg, Andy. 2018. The Untold Story of NotPetya, the Most Devastating Cyberattack in History. *Wired Magazine*, September 2018. https://www.wired.com/story/notpetya-cyberattack-ukraine-russia-code-crashed-the-world/.
Hambling, David. 2006. Techwatch: Forecasting Pain. Popular Mechanics. *Hearst Communications*, New York, 183, no. 12: 32. ISSN 0032-4558.
Hambling, David. 2008. The Microwave Scream Inside your Skull. *Wired Magazine*, 6 July 2008. https://www.wired.com/2008/07/the-microwave-s/.
HelpNet 2019. 43% of Businesses are Still Running Windows 7, Security Threats Remain. Help Net Security, 15 January 2019. https://www.helpnetsecurity.com/2019/01/15/still-running-windows-7/.
Hruska, Joel. 2019. Microsoft Windows XP Is Finally Dead, Nearly 18 Years Post-Launch. Extreme Tech, 11 April 2019. https://www.extremetech.com/computing/289440-microsoft-xp-is-finally-dead-nearly-18-years-post-launch.
ISPS 1988. International Ship and Port Facility Security (ISPS) Code. 1974/1988. International Maritime Organization, London.
ISPS 2012. Guide to Maritime Security and the ISPS Code. 2012. International Maritime Organization, London. ISBN: 078-92-801-1544-4.
ISWG/MASS 1/2/13. Summary of Results of the First Step of the RSE for SOLAS Chapter XI-2 and the ISPS Code. Finland. Intercessional Working Group on Maritime Autonomous Surface Ships. Agenda item 2, 31 July 2019, para 4.1.1, p. 2.
Kovacs, Eduard. 2018. Hackers can Hijack, Sink Ships – Researchers. *Security Week Magazine*, 8 June 2018. Https://www.securityweek,com/hackers-can-hijacj-sink-ships-researchers.
MSC/Circ. 1154. Guidelines on Training and Certification for Company Security Officers. International Maritime Organization, 23 May 2005.
MSC.1/Circ. 1525. Guidance for the Development of National Maritime Security Legislation. International Maritime Organization, London, 1 June 2016.
MSC-FAL.1/Circ. 3. Guidelines on Maritime Cyber Risk Management. International Maritime Organization, London, 5 July 2017.
National Institute of Standards and Technology (NIST) 2016. Maritime Bulk Liquids Transfer Cybersecurity Framework Profile, NIST, Gaithersburg, MD. 2016.

STCW 1978a. Convention of Standards of Training, Certification and Watchkeeping, 1978, with amendments. Chapter VI. Standards regarding emergency, occupational safety, security, medical care and survival functions. International Maritime Organization.

STCW 1978b. Convention of Standards of Training, Certification and Watchkeeping, 1978. Manila Amendments to the annex to STCW 1978. Mandatory minimum requirements for the issue of certificates of proficiency for ship security officers. Section A–VI/5.

TSA 2019. Transportation Worker Identification Card. TWIC©. Transportation Security Administration. Department of Homeland Security. tsa.gov// for-industry/twic.

USCG 04–19. United States Coast Guard. Marine Safety Information Bulletin. Cyber Adversaries Targeting Commercial Vessels MSIB Number 04-19. Washington, DC, 24 May 2019.

USCG 06–19. United States Coast Guard. Marine Safety Alert. Cyber Incident Exposes Potential Vulnerabilities Onboard Commercial Vessels. Safety Alert 06–19. Washington, DC, 8 July 2019.

USCG 2017. Vessel Security Plan Stage II Checklist. Marine Safety Center. Vessel Security Division. United States Coast Guard, as amended, 27 December 2017. https://www.dco.uscg.mil/Portals/9/MSC/Vessel%20Security/VSP_Stage_ II_Checklist.pdf?ver=2017-12-27-105529-803.

Wright, R. Glenn. 2019. Intelligent Autonomous Ship Navigation using Multi-Sensor Modalities. 15th International Symposium on Marine Navigation and Safety of Sea Transportation (TransNav), Gdynia, Poland, 12 June 2019.

Wright, R. Glenn and Michael Baldauf. 2016. Correlation of Virtual Aids to Navigation to the Physical Environment. *International Journal on Marine Navigation and Safety of Sea Transportation*, 10. no. 2. DOI: 10.12716/ 1001.10.02.11. 287–299.

# Chapter 9

# Training for MASS Operations

Seafarers undergo rigorous academic and practical training to learn and gain experience in their jobs. However, a question exists of whether current training regimes are adequate to cope with the pending digitalization of the maritime industry and the advent of unmanned ships. These are disruptive events in an established industry bringing with them change in the application of new technologies and the methods for their implementation. Such issues spark debate as to whether existing maritime curricula should be trimmed to make room for new material reflecting these changes and, if so, what current material should no longer be taught. If the new material is to be added to existing curricula, will the extra time and effort required to earn maritime degrees and credentials make these occupations uncompetitive with other industries and become out of reach for many? Likewise, the question of how must credentialing be changed to reflect "sea service" and other criteria while stationed at a remote control center (RCC) to remotely operate Maritime Autonomous Surface Ships (MASS) must be considered. This chapter attempts to address these issues and answer some of the many questions that arise on the subject. Hopefully, through this process some light may be shed on practical approaches that may be useful in training new entries into the maritime industry as well as for recurring training of existing maritime professionals to help ensure the integration of MASS into the shipping ecosystem is accomplished in an efficient and painless manner as is possible to introduce this highly disruptive technology.

## 9.1 TECHNOLOGICAL CHANGE

Change is rapidly occurring in both the maritime industry and the people who work there. Through digitalization all aspects of shipping and the companies involved are becoming more interconnected along with the ships they operate. The ships themselves are also becoming much more capable in terms of their equipment and communications complement bringing with it new opportunities to enhance safety of navigation as well as efficiency, and new problems related to cyber security and being the objects of targeted cyber attacks.

Transformation and expansion of education and learning from books, procedural simulators and feet on the deck of operational vessels to include audio, visual and augmented reality formats will be essential to keep up with the rate of change occurring in the maritime industry [Nguyen 2018]. Digitalizing training will also provide new seafarers with a direct and early introduction to the technologies and skills most relevant to their future occupations and help ease the transition of active seafarers into this new realm. Having grown up in a world with readily available Wi-Fi, young adults are generally adept at using smartphones, computer and gaming technology – much more so than those established in the work force for many years who are more familiar with landlines and dial-up connections. As digitalization expands throughout the industry and MASS subsequently gains in market share, maritime education must adapt to train this workforce in the application and use of this new technology yet must still continue to improve teaching processes to complement these new technologies and capabilities by providing perspective with personal insight and experiences.

There are several aspects of the current trend in technological change that apply to the maritime industry in general but more specifically having to do with MASS. This includes the development of new sensor technologies and their maritime use, the implementation of RCCs for remote vessel operation and monitoring, new navigation techniques, enormous increases in data communications requirements with more capable methods, and the use of artificial intelligence for autonomous command decision making. These are all new concepts to the maritime industry as are the technologies that implement them, which represent subjects future seafarers should be acquainted with to various degrees depending upon their specific job functions.

The following paragraphs look at possible future changes to training curricula at maritime academies, universities and academic institutions, and maritime institutes to train deck and engineering officers in these new technologies critical to MASS implementation. Many of these changes are designed to familiarize future and existing seafarers with possible approaches for expanding their knowledge to stay relevant in the era of digitalization and MASS. However, some approaches may open the door for those who traditionally would not pursue a maritime occupation yet may be attracted by the many new professional opportunities that may be presented. Also considered are the impact of credentialing new competencies that may be required under the International Convention on Standards of Training, Certification and Watchkeeping (STCW).

## 9.2 TRAINING CURRICULA

Subjects taught in the education of seafarers are based in part upon centuries of knowledge and experience gained in the traditional skills and customs needed to perform the jobs inherent to the maritime professions. Another part involves providing training in the specific skills proscribed by regulation. This combination of traditional seafarer skills and core competencies that are represented in present-day training curricula is unlikely to be adequate to train future seafarers, considering the effects of digitalization and MASS.

In the United States, six maritime academies produce more than 70% of the new U.S. officers licensed to operate vessels of unlimited tonnage and any horsepower each year [Alfultis et al. 2017]. These graduates have Bachelor of Science degrees and typically hold initial credentials as Third Mates or Engineers. Similar academies exist throughout the world, producing seafarers with equivalent credentials. However, many more seafarers with and without academic degrees have earned and continue to earn maritime credentials of various tonnage ratings by working their way up beginning as unlicensed seafarers and through training received from the many maritime institutes that exist worldwide. A non-traditional source of future seafarers and technical support staff likely to be attracted to MASS operations include people with engineering degrees without maritime credentials who take graduate-level academic instruction, resulting in a Master's degree designed specifically to meet the requirements for MASS. Details are provided on three different approaches representing these scenarios, which may help to address this issue.

### 9.2.1 Bachelor of Science Degree considering Autonomous Shipping

Present-day maritime academies provide extensive education and rigorous practical training where students earn a Bachelor of Science degree along with merchant marine credentials. Such curricula are significantly beyond the usual requirements for a degree alone. Combined with the practical experience acquired through a semester at sea, the amount of time dedicated to the completion of basic maritime education is substantial.

One approach examining curriculum changes needed to be made to educate and train an autonomous ship operator to work in an operations center was examined by the California State University Maritime Academy in Vallejo, California and presented at a forum on MASS [McNie et al. 2019]. This approach was developed through multiple meetings with deck officers and a mechanical engineer with autonomous engineering experience. A question posed in their assessment that has yet to be answered is whether such an operator would need to be a licensed Third Mate. Assumptions made include the existence of a RCC from which MASS operations would be supervised, vessel operation from sea buoy to sea buoy, the potential for the operation of multiple ships, and possession of a Bachelor of Science degree from a four-year institution. Also assumed were Sheridan levels of automation that varied according to the situations involved [Coppin and Legras 2012]. Two scenarios were explored where merchant mariner credentials are and are not required.

Under the first scenario where the operator would have to be a licensed Third Mate, new courses to be added to the curriculum include the subjects of automation systems, mechatronics lab, electrical power systems and lab, and cybersecurity fundamentals: identification, prevention and mitigation. More "sea time" in simulators would also be required. Modification of existing courses would be needed for marine engineering systems, firefighting (with emphasis on fixed systems, strategies and less "hose holding"), ship stability (emergency operations), COLREG in an automated world, and a more applied focus on electricity and electronics.

The second scenario where merchant mariner credentials are not required identified about 30 units (or credits) in the curriculum where courses were of questionable value. These included celestial navigation, cargo vessel operations, marine survival/personal survival craft, port and terminal operations, shipboard medical, Admiralty law, ship operations/marine management lab, swimming, industrial equipment and safety, operational command at sea, and elective courses such as liquefied gas cargoes.

The implications of their study results indicate that approximately 15 new training units (one semester) would be required under either scenario, noting that the first scenario already requires 159 units. Modification of some existing courses would also be necessary. Up to 30 units would also be needed with the second scenario. There were questions about new competencies that may be required under the STCW. A third option was also identified whereby a new kind of merchant mariner credential specifically geared toward autonomous operations may be created. This would focus on necessary competencies while eliminating unnecessary skills. The follow-up to their assessment would involve more discussions with an industry advisory board and other experts.

### 9.2.2 Master Degree with a Focus on Autonomous Shipping

A new curriculum designed for graduate studies under a Master's degree program in Autonomous Maritime Operations has been established by Novia University in Turku, Finland [Novia University 2018]. This program is intended for holders of a University of Applied Sciences (UAS) Bachelor's degree or another appropriate higher education degree with three years of subsequent work experience in a relevant field. Studies comprise 60 European Credit Transfer System (ECTS) credits leading to a Master of Engineering, Autonomous Maritime Operations. This includes six relevant courses plus a thesis, with course titles that include: Introduction to Studies and Marine Operations, Autonomous Vessels and Automation; Artificial Intelligence, Machine Learning and Human-Machine Interaction; Remote Operations, Cyber Security and Connectivity; and Classification, Qualification and Safety Perspectives.

Novia UAS and the Aboa Mare Maritime Academy and Training Center cooperate with several universities, both nationally and internationally, as well as maritime companies to provide many diverse opportunities for training and consulting services. Graduates of this program are expected to be better suited for employment related to digitalization in the maritime sector and MASS.

### 9.2.3 Maritime Training Centers

Many training centers exist worldwide that provide nationally approved deck and engineering courses essential to assist new entries into the maritime professions as well as to help active maritime professionals to advance their careers. Instruction is generally geared toward preparing for national and STCW Deck and Engineering license exams and includes classroom training and supplemental, hands-on training in simulators to reinforce the principles learned in the classroom. The scope of their training class offerings can span 100 or more subjects with instruction strategies that can be tailored to the needs of small groups and individuals to assist in meeting their personal goals in rapid fashion. Classes are often repeated frequently throughout the course of the year without having to stick to a semester-by-semester schedule. These centers are often the most expedient means for seafarers to acquire the training necessary for their next endorsement or rating.

There are at present no existing, MASS-related endorsements or ratings required for either national or STCW maritime licensing. Maritime training centers as described in this paragraph will generally update existing courses and adopt new curricula needed to fulfill course and licensing requirements as changes in regulations dictate. However, new requirements pertaining to MASS are still likely to be several years in the making with a phase-in period to occur much like that which existed for the introduction of Electronic Chart Display and Information System (ECDIS).

## 9.3 LICENSING REQUIREMENTS

The purpose of the regulatory scoping exercise performed by the International Maritime Organization (discussed in Chapters 10 and 11) concluding in May 2020 at the 102nd meeting of the Maritime Safety Committee (MSC102) determined what specific regulatory instruments: 1) apply to MASS and prevent MASS operations; 2) apply but do not prevent

MASS operations and no action is required; 3) apply but do not prevent MASS operations but need to be amended, clarified or may contain gaps; and 4) have no application to MASS. The second phase beginning in 2020 involves assessing the scope of change required to these regulatory instruments and the directions in which these changes may take place. Specific decisions regarding credentialing seafarers for services related to MASS will result from the outcome of this second phase. Notwithstanding the course of events coming out of future International Maritime Organization (IMO) deliberations, speculation is provided on possible approaches that may be considered.

Certain assumptions regarding MASS crew requirements as discussed in Chapter 5 can be made as a starting point for discussions and include:

- Licensed deck officers will be necessary to command MASS and perform duties as Officer in Charge of a Navigation Watch,
- Unlicensed ratings will be necessary to perform duties as Rating Forming Part of the Navigational Watch for Able Seafarers,
- Licensed engineering officers will be necessary to operate and maintain propulsion plants and related support systems,
- Vessel Security Officer (VSO) functions are required for MASS as for any other vessel, but must be extended to meet the unique needs for MASS,
- A new requirement for Communications Officer is needed to oversee and maintain all means of data and information communications via satellite and radio for MASS, and
- A new requirement for Drone Operator with specific maritime knowledge and experience is needed to safely operate MASS-based remote and autonomous undersea vehicles (RUV/AUV), surface vehicles (RSV/ASV) and aerial vehicles (RAV/AAV).

For MASS operating under autonomy IMO Degree Three Autonomy (remote control) and IMO Degree Four Autonomy (fully autonomous) conditions, these positions would be filled at a RCC or other operating location, which may be on another vessel.

Licensed deck officer endorsements and unlicensed ratings have different requirements for sea service, assessments and approved training, none of which presently reflect any specific requirements for MASS. The following paragraphs provide some thoughts on specific training and experience requirements that may be considered as the IMO proceeds into the second phase of the regulatory scoping exercise for MASS.

### 9.3.1 Licensing Endorsements

National and STCW deck and engineering officer endorsements stipulate specific sea service requirements for their occupations. For deck officers this is specified in terms of the numbers of "days of service" needed to qualify as Master and Mate, with notes as to geographical (i.e., inland, near coastal, and oceans) and tonnage limitations that may exist in the endorsement. Given the assumption that licensed deck officers will be necessary to command MASS and perform duties as Officer in Charge of a Navigation Watch, there is nothing specific to MASS that should drive a need to change any of the presently existing days of service requirements. However, a new requirement for days of service specific to MASS is needed to ensure sufficient familiarity with the control systems, communications and remote control center operations that are unique to MASS. The same rationale exists for national and STCW engineering officer endorsements, where Chief and Second Engineer will be necessary to operate and maintain propulsion plants and related support systems and notes exist citing power rating limitations in terms of kilowatts and horsepower. New licensing requirements should exist not so much for the nature of propulsion systems, but how they interact with overall MASS vessel infrastructure and automated reasoning system control. Sea service requirements for deck and engineering officers seeking MASS endorsements should include the completion of 90 days located at a RCC, either on land or on another vessel, and being directly involved in the actual control and digital twin simulation of MASS appropriate to their tonnage limitations. Should RCC experience be limited to significantly lower tonnages, this lower tonnage should be cited in the endorsement and corresponding MASS operations limited to this lower amount.

Assessment of competence for operating MASS should also be required through either demonstration or simulation or a combination of both. However, appropriate standards would need to be developed. The inclusion of a requirement to show additional training in terms of a class and examination in MASS Operations should also be considered.

### 9.3.2 Unlicensed Ratings

Unlicensed ratings for STCW Able Seafarer and national Able Seaman also stipulate specific sea service, assessment and training or course requirements that will need to be updated, along with the STCW Rating Forming Part of a Navigational Watch (RFPNW) to reflect the unique requirements to stand watch on MASS. This should be sufficient to ensure familiarity with the various sensor systems and displays present at the helm (and possibly multiple ships' helms) as well as control systems, communications and remote control center operations unique to MASS. Sea service requirements for STCW Able Seafarer seeking MASS endorsements should include the completion of 90 days located at a RCC, either on land or on another vessel, and being directly involved in the actual control and digital twin simulation of MASS.

Assessment of competence for operating MASS should also be required through either demonstration or simulation or a combination of both. However, as for licensed endorsements, appropriate standards would need to be developed. The inclusion of a requirement to show additional training in terms of a class and examination in MASS Operations should also be considered.

### 9.3.3 Vessel Security Officer (VSO)

Existing requirements for approved sea service for VSO appear adequate and do not need to be changed. However, additional requirements specific to meet the needs of MASS security should be considered. For example, 360 days of approved sea service on any of several specific vessel types is presently acceptable in fulfilling sea service requirements, yet MASS is not one of the stated vessel types included in this description. The alternative existing requirements for 180 days of approved service and evidence of knowledge or training in basic vessel layout and construction, shipboard organization, shipboard safety, protection of the marine environment and familiarity with key definitions, terminology and operational practices employed in the maritime industry also do not address unique requirements specific to MASS. A possible solution to the sea service issue is to add MASS as one of the approved vessel types and to include MASS-specific knowledge or training.

### 9.3.4 Communications Officer

Current and previous generations of ships where onboard seafarers command and operate the vessel using all methods and tools at their disposal are not limited in their abilities to safely operate and navigate should a failure of communications occur. This same assumption may possibly be made for a future IMO Degree 4, fully autonomous MASS operating on its own without direct participation on the part of a human operator located at a RCC. However, this assumption is entirely premature and as yet unproven, nor can it be proven until sufficient experience in the design, implementation and operation of MASS has been attained. For the foreseeable future, the line of communications between MASS and the RCC will be indispensable and every effort must be made to ensure that communications must be maintained and secure and cannot be interrupted. Should interruption occur qualified personnel must be in charge to ensure it is rapidly reinstated in a proper and complete manner and that adequate plans exist for recovery with as little disruption as possible.

For these reasons a new endorsement for Communications Officer or equivalent deck officer position should be considered. Sea service requirements should be similar to that of the VSO and include 360 days of approved service on MASS with alternative requirements for 180 days of approved service and evidence of knowledge or training in internal vessel communications networks, modes of communication including radio telemetry; conventional and emergency MF/HF, VHF and UHF communications; and Bluetooth, Wi-Fi and wide area networks, cellular and satellite networks. Additional topics should include an overview of shipboard communications system layout and installations, shipboard safety, and familiarity with key definitions, terminology and operational practices employed in MASS. Also note that as the importance and significance of data communications and bandwidth expands to conventional manned vessels, and data exchange between ships and their operations centers greatly increases in the future, this new deck officer endorsement may also need to be extended to cover all vessels, manned and unmanned.

### 9.3.5 Automation Officer

Although fairly new to the maritime industry, the use of machine learning and artificial intelligence-based systems is being incorporated into the Deck and Engineering departments on individual systems over the last several years. However, these technologies will form the mainstream of technology around which MASS operations will be centered. A new endorsement for Automation Officer or equivalent deck officer position should be considered. Sea service requirements should be similar to that of the Vessel Security Officer and Communications Officer and include 360 days of approved service on MASS with alternative requirements for 180 days of approved service and evidence of knowledge or training in reasoning system technology during actual operation as well as during simulations. Knowledge is required of machine learning methods, neural network architectures, training sets used for development, testing, verification and validation of these systems. In addition, demonstration should be required of the ability to perform and supervise digital twin simulations conducted in real time using cloud computing or at the RCC itself using the same exact sensor and other data as the onboard automated reasoning system to ensure similar and functionally equivalent results are attained by both systems. Knowledge and experience should also be demonstrated of causal criteria for reasoning system results; establishment of rationale for their occurrence; and actions required to resolve problems and to perform follow-up development, testing and installation of automated reasoning software. As the importance and significance of artificial intelligence expands to conventional manned vessels in the future, this new deck officer endorsement may also need to be extended to cover all vessels, manned and unmanned.

### 9.3.6 Maritime Drone Operator

Many countries require the licensing of remote drone aircraft operators to ensure their safe operation and that the knowledge of operators is adequate as to where flight may be allowed and/or is specifically prohibited. However, a case can be made that the maritime operating environment is sufficiently unique that drones, including remotely operated and autonomous undersea, surface and aerial vehicles (RUV/AUV, RSV/USV and RAV/AUV), that operate from a vessel should require a new deck officer endorsement such as Maritime Drone Operator to ensure overall safety not only for MASS but for other vessels, people and infrastructure that may exist in the area of their use.

Sea service requirements should be in addition to national licensing requirements and include 90 days of approved services on MASS with alternative requirements for 45 days of approved service and evidence of knowledge or training in maritime specific subjects that include overall safety, shipboard launch and recovery procedures, minimum clearance requirements from people and other vessels, buildings and structures adjoining waterways, communications requirements, and familiarity with key definitions, terminology and operational practices employed in MASS as may be applicable to drone operations. Also note that as the importance and significance of drone operations expands to conventional manned vessels this new deck officer endorsement may also need to be extended to cover all vessels, manned and unmanned.

### REFERENCES

Alfultis, M., E. Fink and M. Wolley. 2017. State Maritime Academies: Educating the Future Maritime Workforce. *The Coast Guard Journal of Safety and Security at Sea. Proceedings of the Marine Safety and Security Council.* United States Coast Guard, January–April 2017.

Coppin, Gilles and Legras, Francois. 2012. Autonomy Spectrum and Performance Perception Issues in Swarm Supervisory Control. Figure 7. *Proceedings of the IEEE*, 100, no. 3: 590–602. DOI: 10.1109/JPROC.2011.2174103.

McNie, E., T. Burback, D. Weinstock, S. Browne and M. Holden. 2019. Education and Training for Autonomous Shipping: Implications for a Maritime University. Presentation for Critical Mass-MARAD. Linthicum, MD, 22–23 July 2019.

Nguyen, Lili. 2018. Digital Training in the Maritime Industry. KNect365 (Informa Connect), 15 March 2018. https://knect365.com/shipping/article/01c65f13-9121-477f-8a21-4beeaf69dba2/digital-training-in-the-maritime-industry.

Novia 2018. Novia University 2018. Master of Engineering, Autonomous Maritime Operations. Turku, Finland. https://www.novia.fi/studies/master-degree-programmes/master-of-engineering-autonomous-maritime-operations/.

# Chapter 10

# Regulatory Issues

The maritime industry features many authorities at the international, regional and national levels as well as classification societies and non-governmental organizations that aid in furthering progress in the development of unmanned and autonomous ships. Their efforts result in regulations, standards, practices, guidelines and recommendations that can promote and enhance the construction, outfitting of vessels and training of those involved in operating and supporting these vessels both on board and shore side.

Leading efforts to coordinate regulatory issues at the international level is the International Maritime Organization (IMO). In 2017, the Maritime Safety Committee (MSC) established the Working Group on Maritime Autonomous Surface Ships (MASS) to perform a regulatory scoping exercise regarding the use of such vessels. This includes the development of guidelines for MASS trials as well as definitions for MASS and the varying degrees to which they may operate independently from human interaction.

Representatives of the various nation states, classification societies and non-governmental organizations (NGOs) participate in MSC activities by providing information, writing and reviewing comments, documents, draft and final regulations along with revisions to regulations. Many have begun work in their respective areas of expertise while others have yet to consider the various issues involved in furthering the framework under which the planning, implementation and operation of such vessels must occur. The breadth of issues considered span vessel equipage and carriage requirements, the many supporting technologies, infrastructure requirements and human staffing needs both aboard ship and on shore to ensure success and safety. An overview of many such authorities and the organizations that help promote regulatory issues for unmanned and autonomous ships is shown in Figure 10.1. This list is not all-inclusive, but it is representative of many that are active in the industry and at the IMO.

This chapter discusses who is involved in MASS regulatory efforts in terms of organizations and the results of their efforts. However, the actual regulatory instruments considered and their purposes are discussed in Chapter 11 – "Legal Issues".

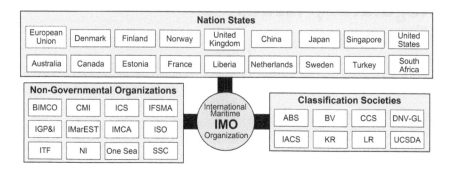

*Figure 10.1* Overview of contributing organizations for the MASS regulatory scoping exercise.

## 10.1 INTERNATIONAL MARITIME ORGANIZATION (IMO)

International efforts lead by the International Maritime Organization pertaining to MASS development, implementation, operation and related standards that may result in future regulation range from ship construction, to equipment carriage requirements, training and other factors. The IMO acts on behalf of maritime operations on technical issues at the international level to promote the adoption of conventions, protocols, codes and recommendations concerning maritime safety and security, the prevention of pollution and related matters [IMO 2013]. With most of its work carried out in committees and sub-committees, the MSC is the main organ that deals with all matters relating to the safety of shipping as well as addressing maritime security issues and piracy and armed robbery against ships.

In June 2017 at its 98th session (MSC 98) the MSC established a new item in their agenda to amend the regulatory framework to enable the safe, secure and environmental operation of partly or entirely unmanned MASS and their interaction and co-existence with manned ships within the existing IMO instruments [MSC 98/20/2]. This included a regulatory scoping exercise regarding MASS with the aim of identifying IMO regulations which, as currently drafted, preclude unmanned operations; regulations that would have no application to unmanned operations as may relate purely to a human presence on board; and regulations which do not preclude unmanned operations but may need to be amended to ensure that the construction and operation of MASS are carried out safely, securely and in an environmentally sound manner.

The scope of this effort included agreement to a high-level definition of MASS and degrees of automation, from partially automated systems to assist crews in the performance of their duties to fully automated ships capable of performing all aspects of ships' operations without need for human intervention. Issues considered include remote control and autonomous navigation, vessel monitoring and collision avoidance systems. Indirectly considered is how human roles fit in with interaction on board other vessels and in shore-based roles. The IMO approved a plan of work, developed a preliminary methodology and established a correspondence group to test the methodology considering existing International Convention for the Safety of Life at Sea (SOLAS) and International Convention of Load Lines (LL) regulations in preparation for full commercial utilization within a decade.

During MSC 99 in June 2018 a preliminary framework for a regulatory scoping exercise pertaining to the use of MASS was approved with a target completion date of 2020 [MSC 99/5]. This framework was carried out at a high level and included objectives, definitions, instruments to be considered, methodology and a plan of work. A cross-divisional MASS taskforce was also established to coordinate work efforts between different IMO bodies [LEG 105/11/1]. The report of the Working Group on MASS was considered and approved by the MSC [MSC 99/22a]. This endorsed the framework for the regulatory scoping exercise including the aim and objective, a preliminary definition of MASS and degrees of autonomy, the list of mandatory instruments to be considered, the applicability in terms of type and size of ships, the methodology for the exercise and a plan of work. A correspondence group (CG) was established to test the framework, make suggestions for improvement and report their results to MSC 100. In particular the CG conducted an initial consideration of SOLAS regulations II-1/3-4 ("Emergency towing procedures for all ships"), III/17-1 ("Recovery of persons from the water"), V/19.2 ("Shipborne navigational systems and equipment") and V/22 ("Navigation bridge visibility"), along with Load Line regulation 10 ("Information to be supplied to the master"). Interested member states and international organizations were also invited to submit proposals to MSC 100 related to the development of interim guidelines for MASS trials.

The report of the CG on MASS developed by Finland was considered at MSC 100 in December 2018 along with comments on this report provided by the Secretariat and several of the member states [MSC 100/5, 100/5/4-8]. Participating in this CG were 33 member states, 2 associate members, 2 observers and 15 NGOs that produced comments on approaches for the regulation of autonomous systems in terms of assessing IMO instruments to see how they may apply to ships with varying degrees of autonomy. Following testing of the methodology by a correspondence group, the MSC approved the framework and methodology for the regulatory scoping exercise [MSC 100.Sum]. The International Organization for Standardization (ISO) contributed an example of a possible characterization scheme for degrees of MASS autonomy and described their plans to launch a new work item to provide an industry standard for basic terminology and concepts for MASS [MSC 100/5/1]. Norway and Baltic and International Maritime

Council (BIMCO), and Korea, submitted guidelines to assist operators and users of MASS test areas and trials in defining services and rules for the use of test areas and trials [MSC 100/5/2, 3]. Also submitted was an initial review of IMO instruments under the purview of MSC by the Secretariat, a preliminary analysis of the International Regulations for Preventing Collisions at Sea (COLREG) by China and the results of technology assessment on MASS by Korea [MSC 100-INF.3, 6 and 10].

It was determined that for each instrument related to maritime safety and security, and for each degree of autonomy, provisions will be identified which apply to:

1. MASS and prevent MASS operations; or
2. MASS and do not prevent MASS operations and require no actions; or
3. MASS and do not prevent MASS operations but may need to be amended or clarified, and/or may contain gaps; or
4. have no application to MASS operations. [MSC 100.Sum]

The degrees of autonomy identified in Chapter 3 were used for the purpose of the scoping exercise. With this in mind the first step of this exercise focused on identifying, in the relevant treaties, provisions that pertain to the above four issues. Upon completion of this first step, the second step is to analyse and determine the most appropriate way of addressing MASS operations, taking into account issues that include the human element, technology and operational factors. The analysis includes the need for:

- Equivalences as provided for by the instruments or developing interpretations, and/or
- Amending existing instruments, and/or
- Developing new instruments, or
- None of the above as a result of the analysis.

### 10.1.1 Regulatory Scoping Exercise

The initial review of instruments under the purview of the Maritime Safety Committee was conducted during the first half of 2019 by a number of volunteering member states with the support of interested international organizations. At the MSC 101 meeting in June 2019 progress was made with the scoping exercise to look at how the safe, secure and environmentally sound operation of MASS may be introduced in IMO instruments [MSC 101.Sum]. The intersessional working group (meeting in September 2019) was tasked with considering the results of the first step and considering how the outcome of the second step reported to MSC 102 at the meeting in May 2020. This report should be based on a high-level discussion on the gaps, themes and/or relevant findings identified during the first step, providing guidance to member states for use in the second step, and providing a report to MSC 102.

## 10.1.2 Interim Guidelines for MASS Trials

The Committee also approved interim guidelines for MASS trials. Among other things, the guidelines say that trials should be conducted in a manner that provides at least the same degree of safety, security and protection of the environment as provided by the relevant instruments. Risks associated with the trials should be appropriately identified and measures to reduce the risks, to as low as reasonably practicable and acceptable, should be put in place. Onboard or remote operators of MASS should be appropriately qualified for operating MASS subject to the trial. Any personnel involved in MASS trials, whether remote or onboard, should be appropriately qualified and experienced to safely conduct MASS trials. Steps should be taken to ensure sufficient cyber risk management of the systems and infrastructure used when conducting MASS trials [MSC 101/WP.8].

The actual regulations to be applied specifically to MASS, including new as well as modifications and exemptions (if any) to existing regulations, are yet to be determined. However, the foresight afforded nation states and the maritime industry by the regulatory scoping exercise through anticipation of regulatory requirements and planning to fulfill them in advance of actual need can help to avert accidents and setbacks that can impede progress in achieving MASS goals and the technologies that support their development.

## 10.2 NATION STATES

Regulatory efforts pertaining to unmanned and autonomous ships performed by nation states through participation at IMO and within their own borders is being accomplished by the European Union, the United Kingdom, China, Japan, Singapore, the United States and other countries in the development of the ships themselves, the technologies needed to operate them and the establishment of locations where test and evaluation may be accomplished. Several countries including the United States are also heavily involved in the development of autonomous ship technologies for use in naval warfare that are outside the scope of this book. However, many civilian initiatives mirror these military technologies and dual use applications are anticipated. These nations participate in IMO activities and directly promote research and development of MASS and their supporting technologies by funding innovative projects, establishing testbeds and transitioning experimental systems in practical applications.

## 10.2.1 European Union

The European Commissions under its Seventh Framework Programme co-funded the Maritime Unmanned Navigation through Intelligence in Networks (MUNIN) project as a collaborative research project to develop and verify a concept for an autonomous ship, defined as a vessel primarily guided by automated onboard decision systems but controlled by a remote operator in a shore side control station [MUNIN 2016]. Conclusions reached from the research include unmanned and autonomous ships can and will be applied where they will be both safer and more cost-effective, and new buildings of autonomous ships are preferred rather than retrofitting existing ships. A list of critical design factors characterizing a viable autonomous and unmanned ship concept was also developed.

The European Research Commission under the Horizon 2020 smart, green and integrated transport work program is funding autonomous ship development projects with a focus on inland waterways, short sea shipping, ferries coastal operations and urban water transport. The goal is to develop and demonstrate a fully autonomous vessel within a realistic environment which encompasses all necessary features including collision avoidance, interaction with waterway and/or port infrastructure, interaction with waterborne traffic, connectivity, control, navigation and docking, condition monitoring, smart maintenance and fail safe operation [ERC MG322018].

## 10.2.2 Denmark

The Danish Maritime Authority (DMA) has been very active in technological development and autonomous maritime solutions for defining a framework to overcome technical and regulatory barriers. DMA sponsored a pre-investigation report prepared by the Technical University of Denmark to describe the potential for autonomous ships and how to define various levels of autonomy [Blanke et al. 2016]. This report describes the lowest autonomy level as having completely manual operation where the navigating officer gets his information from electronic charts about position, course and speed as well as an overview from radar that also presents other ships' course and speed. Autonomy levels progress over various levels of decision-support where automation increasingly performs more tasks, to full autonomy that requires no human participation.

DMA also performed an analysis of regulatory barriers to autonomous ships in civilian commercial shipping that explores Danish law, EU regulation and IMO conventions to conclude the regulatory approach to autonomous shipping should be considered carefully to prevent regulation from becoming a hindrance to technological developments and the commercial use of autonomous technologies in shipping [DMA 2017].

They further conclude that test results from autonomous shipping projects should be published to achieve a credible basis for regulation, secure societal support and demonstrate its benefits. Specific recommendations are provided that autonomous ships should be incorporated into the existing regulatory frame and new regulation should cover areas unique to autonomous ships that existing regulation does not consider. It also re-examines the definition of the term of a master and explores the granting of permits for periodic unmanned navigation bridge and electronic lookout.

DMA has also sponsored a joint project on MASS between CORE Advokatfirma and Nordic Association of Marine Insurers (CEFOR) focusing on identifying the challenges, elements of uncertainty and stakeholders for the introduction of MASS from a civil liability and insurance perspective. This resulted in the publication of their results in December 2018 [CORE/CEFOR 2018].

Denmark provided a preliminary analysis of regulations to aid the IMO MASS regulatory scoping exercise, including related instruments and regulations and regulatory barriers and recommendations [MSC 99/INF.3].

## 10.2.3 Finland

Tekes, the Finnish Funding Agency for Technology and Innovation, is funding collaborative research on autonomous ships under the Arctic Seas program in partnership with Rolls-Royce and other industrial firms with a goal of achieving autonomous merchant vessels by 2015 in creating an autonomous maritime ecosystem [Tekes 2016]. This partnership forms the One Sea Autonomous Maritime Ecosystem where Big Data intelligent devices, artificial intelligence and remote control technologies coalesce toward increases in reliability and operational efficiency through digitalization by adopting intelligent solutions to shipping on a large scale [One Sea 2018]. This is planned to be accomplished partly through retrofit of new technologies into older ships and the development of a new fleet of smart ships. Initial testing of autonomous ship applications is expected to be accomplished by 2020. The Jaakomeri test area has been established in the coastal area of Finland that is open to all companies, research institutes and others for full-scale tests of autonomous ships, maritime traffic concepts, and related technologies [MSC 99/INF.13]. Funding is provided by the Finnish Government Bureau, Centre for Economic Development, Transport and the Environment of Southwest Finland and the test area is managed and controlled by DIMECC Ltd.

Finland provided an analysis of definitions for different concepts and levels of autonomy suggested by the industry (Bureau Veritas, Lloyd's Register, the Norwegian Forum for Autonomous Ships (NFAS), Ramboll-Core, Rolls-Royce, UK Marine Industries Alliance) [MSC 99/5/6].

### 10.2.4 The Netherlands

The Industry Project Autonomous Shipping is a joint two-year applied research and development project to explore possible applications and to study the requirements for safe navigation in a shipping environment. It consists of a consortium of 17 partners including maritime businesses and shipping companies, academia and the Dutch government represented by the Ministry of Infrastructure and Water Management and the Ministry of Defence [JIPAS 2019]. It is partly funded by the TKI-Maritiem allowance of the Dutch Ministry of Economic Affairs and Climate Policy [MAREX 2019].

### 10.2.5 Norway

The Norwegian Forum for Autonomous Ships (NFAS) was formed through collaboration with the Norwegian Maritime Authority (NMA), Norwegian Coastal Administration (NCA), the Federation of Norwegian Industries and SINTEF Ocean to examine autonomous vessels considering public acceptance, acceptable risk levels, efficient design and approval processes and other challenges [NFAS 2018]. NFAS goals include promoting cooperation between users, researchers, authorities and others interested in autonomous ships and their use; contributing to the development of common Norwegian strategies for development and use of autonomous ships; being a common voice for the development and use of autonomous ships based on the interest organizations objectives; and strengthening the Norwegian interest group's international contacts and influence within the area of autonomous ships. This is accomplished through sponsoring many conferences and meetings throughout the year where NFAS members and others involved with and interests in autonomous ships may participate.

The NCA has designated official test areas for autonomous shipping at locations in Trondheimsfjord, Horten and Storfjorden. Their uses include the development of technology for autonomous ships as well as conducting autonomous ship trials. Participating in these test area initiatives are Det Norske Veritas-Germanischer Lloyd (DNV-GL), the Norwegian Defence Research Establishment (FFI) and the University College of South East Norway, Norwegian Marine Technology Research Institute (MARINTEK), the Norwegian University of Science and Technology (NTNU) Centre for Autonomous Operations and Services (AMOS), the Trondheim Port Authority, Kongsberg, Maritime Robotics and the Ocean Space Centre.

At MSC 99 Norway supplied a presentation on the development of the autonomous containership *Yara Birkeland* [MSC 99/INF.16].

### 10.2.6 United Kingdom

The United Kingdom launched the Maritime 2050 initiative which is viewed as an important component of a suite of strategy and policy documents that will together establish the framework for marine and maritime matters in the United Kingdom looking forward the next 30 years [Maritime UK, 2018]. An Industry Code of Practice was also released by the UK Maritime Autonomous Systems Regulatory Working Group (MASRWG) in November 2017 providing new guidance focusing on operations, skills, training and vessel registration for MASS [Maritime UK, 2017]. Developments in autonomous ship technology include the multipurpose work-class vessel *C-Worker7* for a variety of offshore and coastal engineering, and the unmanned and remotely controlled vessel *RALamander* for fire-fighting in ports or at sea [Wingrove 2018].

## 10.2.7 China

China has started the development of a 225-square-mile test area for autonomous ships in Zhuhai, Guangdong. The test area is expected to be the country's main base for research into autonomous ship technology, including obstacle avoidance technology, over the coming three to five years. The project is being undertaken by the Zhuhai government, China Classification Society (CCS), Wuhan University of Technology and Zhuhai Yunzhou Intelligence Technology [Ship Technology 2018].

At MSC 99 China and Finland contributed comments on the challenges faced by MASS and the role of IMO. It has also proposed a work plan with deliverables for the regulatory scoping exercise, the consideration of a goal-based approach, the development of interim guidelines for MASS trials on international voyages and the establishment of a mechanism for information-sharing [MSC 99/5/7]. China and Liberia also provided recommendations to consider manned MASS (with crew on board) and unmanned MASS (without crew on board) along with the adoption of risk assessment methods when carrying out the regulatory scoping exercise. Recommendations were also provided to prioritize the development of interim safety and environmental protection guidelines for unmanned cargo carriers [MSC 99/5/8].

## 10.2.8 Singapore

The Maritime Port Authority of Singapore (MPA) has renewed a Memorandum of Understanding (MOU) with Det Norske Veritas/ Germanischer Lloyd (DNV-GL) to jointly develop autonomous vessels [MPA 2018]. MPA is also partnering with Keppel Offshore & Marine (Keppel O&M) and the Technology Centre for Offshore and Marine Singapore (TCOMS) to jointly develop autonomous vessels capable of undertaking harbor operations such as channeling, berthing, mooring and towing, for safer, more efficient and more cost-effective operations. Under this project a digital twin of the tug will also be developed to simulate vessel behavior to optimize vessel operations using data analytics and visualization tools [MPA 2018-1, 2018-2].

### 10.2.9 Japan

Several shipbuilders and shipping firms in Japan are jointly developing remote-controlled cargo vessels with a goal to launch by 2025 [*Nikkei Asian Review* 2017]. The ships would use the Internet of Things (IoT) to connect a range of devices over the Internet to gather data including weather conditions and shipping information and plot the shortest, most efficient and safest routes. One goal is to remove the potential for human error to dramatically cut the number of accidents at sea. Mitsui OSK Lines, Nippon Yusen and other firms participating in this project anticipate investing hundreds of millions of dollars developing the technology required to steer as many as 250 ships through busy shipping lanes.

Japan's Ministry of Land, Infrastructure, Transportation and Tourism (MLIT) has selected a joint demonstration project proposed by Mitsui OSK Lines (MOL) on autonomous shipping [MLIT 2018]. It has been selected for MLIT's FY2018 autonomous vessel demonstration project with an aim to bring vessels with autonomous berthing and un-berthing into service by 2025.

At MSC 99 Japan provided comments on the regulatory scoping exercise, in particular on the importance of the recognition of a phased development of new and advancing technologies covering several phases between conventional ships and unmanned operations. These comments highlighted the importance of considering safety requirements for new and advancing technologies which could be introduced gradually leading toward MASS and forming a common understanding to lead to lengthy debates such as manned MASS. A list of tasks was prepared for relevant sub-committees and terms of reference for a cross-committee Working Group on MASS and the development of guiding principles to underpin the Committee's approach [MSC 99/5/9]. Japan also provided three sets of outcomes of studies conducted in Japan on mandatory regulations under the International Convention for the Safety of Life at Sea (SOLAS), International Convention on Standards of Training, Certification and Watchkeeping (STCW) and the International Regulations for Preventing Collisions at Sea (COLREG) relating to MASS [MSC 99/INF.14].

### 10.2.10 United States

Despite having a leading role in IMO activities that includes active participation in MASS functions the United States lags behind much of the rest of the world in assisting the development of commercial autonomous ships. However, the US Coast Guard has been highly supportive in assisting these commercial ventures in testing the technologies and autonomous system testbeds. The US also leads the world in developing naval autonomous ships and their supporting technologies for military applications. Many of these technologies have dual use capabilities and are expected to transition into the commercial marketplace through the participation of U.S. companies in commercial autonomous projects worldwide. One example of such a program is the Defense Advanced Research Projects Agency (DARPA) development and successful demonstration of a highly autonomous unmanned ship, *Sea Hunter*, as part of the Anti-Submarine Warfare (ASW) Continuous Trail Unmanned Vessel (ACTUV) program. This vessel has transitioned to the Office of Naval Research for further development and testing and it is expected to ultimately become an entirely new class of ocean-going vessel able to traverse thousands of miles over open seas for months at a time, without a single crew member aboard [ACTUV 2018]. The US has also announced the development of their medium and large unmanned surface vessel (MUSV/LUSV) concepts and is actively pursuing several projects in these areas.

DARPA is also active in the development of support technologies for autonomous ships through its Assured Autonomy and other initiatives. One example is a DARPA-sponsorship of Michigan Technological University and University of Michigan for their Seaworthiness through Intelligent Trajectory Control and High-Fidelity Environmental Sensing project whose goal is to investigate the ability of autonomous surface vessels to negotiate and survive large sea states and extreme weather conditions without operator intervention [White 2018]. DARPA also sponsors projects that utilize the Marine Autonomy Research Site (MARS) in the Great Lakes.

At MSC 99 the United States provided comments and recommended terms and definitions for different levels of autonomy and possible arrangements and methods of work for the regulatory scoping exercise [MSC 99/5/5, MSC 99/5/12].

## 10.2.11 Other Countries and Related Issues

There are many additional countries involved in the development of MASS initiatives and participating in IMO activities that are not were not previously specifically mentioned. A joint proposal was submitted by Australia, Canada, Denmark, Estonia, Finland, Japan, the Netherlands, Norway, Singapore, Sweden, the United Kingdom and the United States as an approach for the regulatory scoping exercise, including the establishment of working groups and intersessional correspondence groups to complete the output by MSC 102, draft terms of reference for a working group at MSC 99 and expected deliverables [MSC 99/5/5]. Turkey provided comments on several documents and provided recommendations to carry out the regulatory scoping exercise in phases, giving priority to matters less likely to lead to lengthy debates, such as manned MASS, preparing a list of tasks for relevant sub-committees and terms of reference for a cross-committee Working Group on MASS and the development of guiding principles to underpin the Committee's approach [MSC 99/5/11]; France proposed a methodology for the regulatory scoping exercise, as well as definitions for autonomous ships and different levels of autonomy, and two approaches for adapting the regulatory framework [MSC 99/5/4]. Finland, Liberia, Singapore, South Africa and Sweden and other countries provided recommendations on the identification of potential amendments to existing IMO instruments and proposed a two-step approach for the regulatory scoping exercise consisting of identifying and categorizing IMO instruments relevant to the operation of MASS [MSC 99/5/3]. Specific regulations that might require amendments to ensure MASS are operated safely, securely and in an environmentally sound manner were also identified. Comments regarding the interim guidelines for MASS trials were submitted by Finland, Japan, Norway, Republic of Korea, Singapore, United Arab Emirates, BIMCO and the ITF [MSC 101/5/1, 101/5/5].

## 10.3 CLASSIFICATION SOCIETIES

The role of classification societies is to establish and maintain technical standards for the construction and operation of marine vessels and offshore structures. Those actively participating in the MASS regulatory scoping exercise and which are likely to play key roles in the future establishment of MASS are discussed.

### 10.3.1 American Bureau of Shipping (ABS)

The American Bureau of Shipping (ABS) is a leading international classification organization, founded in 1862 and devoted to promoting the security of life and property and preserving the natural environment through the development and verification of standards for the design, construction and operational maintenance of marine and offshore assets [ABS 2018]. In 2017, they joined the Unmanned Cargo Ship Development Alliance, which, along with eight other members, expects to deliver an unmanned cargo ship by October 2021 [ABS 2017]. The design will integrate features of independent decision making, autonomous navigation, environmental perception and remote control.

### 10.3.2 Bureau Veritas (BV)

Created in 1828, Bureau Veritas is a world leader in laboratory testing, inspection and certification services and has around 75,000 employees in more than 1,400 offices worldwide [BV 2018]. In December 2017, BV released their publication "Guidelines for Autonomous Shipping", offering guidance for risk assessment of ships including autonomous systems. Its goal-based recommendations described minimum levels of functionality of autonomous and the guidelines for improving the reliability of essential systems within autonomous ships [BV 2018-1]. It is focused mainly on MASS of 500 GT or more, excluding small ships of less than 20 meters in length and unmanned underwater vehicles [BV-2017].

### 10.3.3 China Classification Society (CCS)

The China Classification Society provides classification services to ships, offshore installations and related industrial products by furnishing world-leading technical rules and standards [CCS 2018]. CCS has jointly established with the HNA Technology & Logistics Group the Unmanned Cargo Ship Development Alliance and is a participant in the construction of a test area for autonomous ships in Zhuhai, Guangdong [WMN 2017, MAREX 2018].

### 10.3.4 Det Norske Veritas-Germanischer Lloyd (DNV-GL)

The merger of Det Norske Veritas and Germanischer Lloyd resulted in Det Norske Veritas-Germanischer Lloyd, a world-leading classification society and a recognized advisor for the maritime industry. Operating in more than 100 countries, they enhance safety, quality, energy efficiency and environmental performance of the global shipping industry. A recently signed Memorandum of Understanding with the Maritime Port Authority of Singapore (MPA) has DNV-GL jointly developing autonomous vessels with the MPA [MPA 2018].

### 10.3.5 International Association of Classification Societies Ltd. (IACS)

The International Association of Classification Societies is a not-for-profit membership organization of classification societies that establishes minimum technical standards and requirements that address maritime safety and environmental protection and ensures their consistent application [IACS 2018]. They announced a series of initiatives designed to foster modernization and pave the way for autonomous shipping and includes a review of IACS Resolutions to identify and remove elements hindering the development of new technologies, including ship autonomy [MAREX 2017]. These include developing procedures relating to the deployment of electronic certificates, while continuing to support the IMO's work in promoting their use throughout the industry. Also included is the modernization of survey methods and enabling the use of new technologies. A summary of work carried out by IACS is provided in their position paper on MASS [IACS 2019].

### 10.3.6 Korean Register of Shipping (KR)

The Korean Register of Shipping works to promote safe ship operations and clean oceans through the development of technology and pursuit of excellence in their rules and standards for shipping, shipbuilding and their industrial services [KR 2018]. They have released their draft document "Guidance for Autonomous Ships" to ensure the safety and reliability of autonomous ships or systems and functions necessary for autonomous operation through risk assessment. It is applicable to construction contracts signed or application for Classification Survey beginning January 2019 [KR 2019].

### 10.3.7 Lloyd's Register of Shipping (LR)

Lloyd's Register of Shipping is a leading international provider of classification, compliance and consultancy services to the marine and offshore industries [Lloyd's 2018]. It has published classification guidance for different levels of ship autonomy. This guidance, identified as the Unmanned Marine Systems Code is intended to provide designers, builders and operators with clarity on identifying the desired level of autonomy [*gCaptain* 2017, Lloyd's 2017]. The first three levels all require that navigational technology exists on a manned vessel to provide support in decision making. The next three levels all involve unmanned vessels with different levels of remote operation, including complete autonomy. Remote operation includes shore-based operators who can intervene when notified by the navigational system. The Code, which has been validated against several existing UMS designs, has been developed against a hazard analysis of unmanned marine system design and operation and benchmarked against existing commercial and naval regulatory requirements, including SOLAS and the Naval Ship Code.

## 10.4 NON-GOVERNMENTAL ORGANIZATIONS

Many non-governmental organizations (NGOs) are active in the development and promotion of MASS and their supporting technologies. These organizations also participate in, provide comments and forwarded a number of proposals to the MSC. Summaries provided are obtained from the MSC Final Reports for their respective meetings as well as from individual documents as may be referenced.

### 10.4.1 Baltic and International Maritime Council (BIMCO)

The Baltic and International Maritime Council, which is the world's largest shipping association, states that it will take a leading position on the facilitation of trading with autonomous ships [BIMCO 2018]. This position includes supporting a set of standard definitions covering the different levels of automation and methods of control to create a clearer framework for future regulation and to develop a common understanding of the risks and opportunities of autonomous ships. It will support the initiatives by the

IMO and CMI to assess the need for changes to international conventions and national laws for autonomous ships to operate worldwide. Finally, it recognizes the importance of the human element, focusing on new competences for seafarers and on the need for human relations initiatives to overcome problems such as potential loneliness following the possible reduction of personnel onboard.

### 10.4.2 Comité Maritime International (CMI)

Comité Maritime International is a non-governmental not-for-profit international organization established in 1897 to contribute by all appropriate means and activities to the unification of maritime law [CMI 2018]. It also promotes the establishment of national associations of maritime law and cooperates with other international organizations. In 2015, the CMI established an International Working Group on MASS to identify the legal issues related to unmanned shipping and to provide an international legal perspective on these issues. It provided a summary of responses received from national maritime law associations to the CMI International Working Group questionnaire on unmanned ships and the work of the group on SOLAS, MARPOL, COLREG, STCW, SAR, SUA, Facilitation Committee and the Salvage Convention [MSC 99/INF.8].

### 10.4.3 International Chamber of Shipping (ICS)

The International Chamber of Shipping (ICS) is the principal international trade association for merchant ship owners and operators, representing all sectors and trades and over 80% of the world merchant fleet [ICS 2018]. At MSC 99 ICS provided comments on autonomous systems and proposed the development of a work plan for the regulatory scoping exercise that should include additional framework tasks to assess the risks and effectiveness of different alternatives for conducting the exercise. This also recommended considering the need for a holistic approach to the regulation of autonomous systems including addressing human element, procedural and technology matters [MSC 99/5/2].

### 10.4.4 International Federation of Shipmasters' Associations (IFSMA)

The International Federation of Shipmasters' Associations representing shipmasters and the International Transport Workers' Federation jointly submitted comments on the regulatory scoping exercise that considered UNCLOS provisions, the involvement of relevant committees and sub-committees and the definition of different types of autonomy. Also addressed was the extent of human monitoring and control and human element components. The possibility of extending the target completion year of the output to 2023 was also considered. A proposal was forwarded that that remotely controlled or unmanned ships should not be permitted to operate on international voyages until an international regulatory framework governing their operation had been adopted [MSC 99/5/1]. The IFSMA and ITF also provided information relating to a survey on autonomous ships representing the views of over 1,000 maritime professionals worldwide [MSC 99/INF.5].

### 10.4.5 International Group of Protection and Indemnity Clubs (IGP&I)

The International Group of Protection and Indemnity Clubs is comprised of 13 P&I Clubs that provide liability cover (protection and indemnity) for 90% of the world's ocean-going tonnage [IGP&I 2018]. They have established an Autonomous Vessel working group and have been active, along with a number of Club members, in consulting with the CMI International Working Group on MASS as well as the joint project on MASS between CORE Advokatfirma and CEFOR sponsored by the Danish Maritime Authority [IGP&I 2018-1].

### 10.4.6 Institute of Marine Engineering, Science & Technology (IMarEST) and the International Marine Contractors Association (IMCA)

The Institute of Marine Engineering, Science & Technology is the international body and learned society for all marine professions that include marine engineers, scientists and technologists [IMarEST 2018]. In September 2019 they published their report on Autonomous Shipping looking ahead to 2050 [IMarEST 2019]. The International Marine Contractors Association is a leading trade association representing the vast majority of contractors and the associated supply chain in the offshore marine construction industry worldwide [IMCA 2018]. IMarEST and IMCA joined with several countries to propose an approach for the regulatory scoping exercise, including the establishment of working groups and intersessional correspondence groups to complete the output by MSC 102 [MSC 99/5/5]. Their contribution also included draft terms of reference for a working group at MSC 99 and expected deliverables.

### 10.4.7 International Organization for Standardization (ISO)

The International Organization for Standardization is an independent international organization with a membership of 162 national standards bodies that brings together experts to share knowledge and develop voluntary, consensus-based, market relevant international standards that support innovation and provide solutions to global challenges [ISO 2018]. ISO Technical Committee 8 (TC 8) for Ships and Marine Technology has established Working Group 10 on Smart Shipping and a task group on MASS. Their focus is on determining how to share the experience of industry for the development of standards supporting intelligent technology and Internet of Things (IoT) for shipbuilding, shipping, ports and logistics.

### 10.4.8 International Transport Workers' Federation (ITF)

The International Transport Workers' Federation representing transport workers provided comments on the regulatory scoping exercise and recommending a phased development to include semi-autonomous systems to support and supplement shipboard functions in conjunction with onboard human supervision and intervention. Sensor and data exchange technology was addressed along with communication links for shore-based monitoring of shipboard functions. Also considered are ships operating in dual mode relying on semi-autonomous systems under routine circumstances with higher levels of onboard human involvement under non-routine circumstances such as navigation in congested waters, rough weather, equipment failure, or unforeseen circumstances, and phasing to shore-based rather than onboard human involvement based on user experience, technical feasibility and cost-benefit analysis [MSC 99/5/10].

### 10.4.9 Nautical Institute (NI)

The Nautical Institute, an international representative body for maritime professionals involved in the control of sea-going ships, has become involved in MASS issues pertaining to the human element of future vessel operations [NI 2017]. They have taken the position that all vessels, regardless of whether they are autonomous or not, should comply with the COLREG [NI 2018].

### 10.4.10 One Sea Autonomous Maritime Ecosystem

The One Sea Autonomous Maritime Ecosystem is a Finnish collaboration with the primary aim toward establishing an operating autonomous maritime ecosystem by 2025. Beginning in 2016, they have gathered together leading marine experts to form a strategic combination of top research, information technology and businesses. Partners include DIMECC, ABB, Business Finland, Cargotec, Ericsson, Finnpilot Pilotage, Inmarsat, Kongsberg, Monohakobi Technology Institute, Tieto and Wärtsilä.

## 10.4.11 Smart Ships Coalition (SSC)

The Smart Ships Coalition (SSC) is a broad community of stakeholders of academic, state and federal agencies, private and non-profit industry, and international organizations with a common interest in the advancement and application of autonomous technologies operated in marine environments in the Great Lakes region and the U.S. coastal oceans [SmartShips 2018]. It has established that the Marine Autonomy Research Site (MARS) Great Lakes test area open to all companies, research institutions, government agencies and others wishing to test autonomous surface and subsurface vehicles and related technologies [SmartShips 2018-1].

## 10.4.12 Unmanned Cargo Ship Development Alliance

The Unmanned Cargo Ship Development Alliance was established in 2017 to promote changes in the ship design and operation as well as facilitating the establishment of technology, regulations and standard systems involved in unmanned cargo ships [CCS 2017]. The Alliance will take advantage of all members' specialties and integrate advanced technology at home and abroad to develop unmanned cargo ships with independent navigational capacity and promote the development of intelligent shipping. It is comprised of shipyards, equipment manufacturers and designers to advance the goals of autonomous shipping [MAREX 2018]. Their goal is to deliver a working autonomous cargo ship by October 2021 that features integrated independent decision making, autonomous navigation, environmental perception and remote control. Attending its first meeting in Shanghai were leaders and industry experts from HNA Technology & Logistics Group, CCS, ABS, China Ship Development and Design Center, Hudong-Zhonghua Shipbuilding (Group) Co., Ltd, China State Shipbuilding Corporation (CSSC), Rolls-Royce, China Shipbuilding Industry Corporation (CSIC) and Wärtsilä.

## REFERENCES

ACTUV 2018. ACTUV "Sea Hunter" Prototype Transitions to Office of Naval Research for Further Development. Defense Advanced Research Projects Agency, 30 January 2018. https://www.darpa.mil/news-events/2018-01-30a.

American Bureau of Shipping (ABS) 2017. ABS Joins Industry Partners to Advance Autonomous Shipping. American Bureau of Shipping Press Release, 26 July 2017. https://ww2.eagle.org/en/news/press-room/ABS-Joins-Industry-Partners-to-Advance-Autonomous-Shipping.html.

American Bureau of Shipping (ABS) 2018. https://ww2.eagle.org/en.html.

Baltic and International Maritime Council (BIMCO) 2018. BIMCO Position on Autonomous Ships. https://www.bimco.org/about-us-and-our-members/bimco-statements.

Blanke, M., M. Henriques and J. Bang. 2016. A Pre-analysis on Autonomous Ships. Technical University of Denmark, DTU Electro, Elektrovej 326 and DTU Management Engineering, Produktionstorvet 426, DK-2800. Kongens Lyngby. www.dtu.dk., Autumn 2016.

Bureau Veritas (BV) 2017. Guidelines for Autonomous Shipping. December 2017, para. 1.2, p. 5.

Bureau Veritas (BV) 2018. Cyber Safety and Autonomous Shipping Addressed with new Bureau Veritas Notations and Guidelines. 13 March 2018. https://group.bureauveritas.com/cyber-safety-security-and-autonomous-shipping-addressed-new-bureau-veritas-notations-and-guidelines.

Bureau Veritas (BV) 2018-1. https://group.bureauveritas.com/proup.

CCS 2017. Unmanned Cargo Ship Development Alliance Launched in Shanghai, 7 December 2017. http://www.ccs.org.cn/ccswzen/font/fontAction!article.do?articleId=4028e3d65d11c4cb015d35ede4af002c.

China Classification Society (CCS) 2018. http://www.ccs.org.cn/ccswzen/font/fontAction!moudleIndex.do?moudleId=297e62d739e7b9c0139ebcf090b0004.

Comite Maritime International (CMI), 2018. https://comitemaritime.org/about-us/.

CORE/CEFOR 2018. Maritime Autonomous Surface Ships, Zooming in on Civil Liability and Insurance. CORE Advokatfirma and Cefor, December 2018. https://static.mycoracle.com/igpi_website/media/article_attachments/maritime_autonomous_ship_report_dec_18_6URrRcy.pdf.

Danish Maritime Authority (DMA) 2017. Analysis of Regulatory Barriers to the Use of Autonomous Ships. Final Report. Prepared by Rambøll and the CORE Law firm, December 2017.

ERC MG322018. European Research Commission. Topic: MG-3-2-2018, 27 October 2017. https://ec.europa.eu/research/participants/portal/desktop/en/opportunities/h2020/topics/mg-3-2-2018.html.

*gCaptain* 2017. Lloyd's Register Announces New Code to Certify Unmanned Ships. 14 June 2017. http://gcaptain.com/lloyds-register-announces-new-code-to-certify-unmanned-ships/.

IGP&I 2018-1. Autonomous Ships. International Group of P&I Clubs, 14 December 2018. https://www. igpandi.org/article/autonomous-ships.

Institute of Marine Engineering, Science & Technology (IMarEST) 2018. https://www.imarest.org/about-imarest.

Institute of Marine Engineering, Science & Technology (IMarEST) 2019. Autonomous Shipping: Putting the Human Back in the Headlines. Manila, Philippines, September 2019.

International Association of Classification Societies (IACS), 2018. www.iacs.org.uk/about/.

International Association of Classification Societies (IACS) 2019. Position Paper: MASS. www. iacs.org.uk/media/5926/iacs-mass-position-paper.pdf.

International Chamber of Shipping (ICS) 2018. www.ics-shipping.org.

International Group of P&I Clubs (IGP&I) 2018. https://www.igpandi.org/about.

International Marine Contractors Association (IMCA) 2018. www.imca-int.com/about-imca/.

International Maritime Organization (IMO) 2013. IMO, What It Is. October 2013, pp. 2, 4. http://www.imo.org/en/About/Documents/What%20it%20is%20Oct%202013_Web.pdf.

Joint Industry Project Automated Shipping (JIPAS) 2019. www.autonomousshipping.nl.
Korean Register of Shipping (KR) 2018. www.krs.co.kr/eng/main.aspx.
Korean Register of Shipping (KR) 2019. Guidance for Autonomous Ships. GC-28-E KR, para. 102, p. 1.
LEG 105/11/1. Proposal for a Regulatory Scoping Exercise and Gap Analysis with Respect to Maritime Autonomous Surface Ships (MASS). IMO Legal Committee, 19 January 2018. Canada, Finland, Georgia, the Marshall Islands, Norway, the Republic of Korea, Turkey, CMI, ICS and P&I Clubs.
Lloyd's 2017. LR Code for Unmanned Marine Systems. Lloyd's Register, February 2017.
Lloyd's 2018. Lloyd's Register. https://www.lr.org/en/marine-shipping.
MAREX 2017. IACS Chairman Prioritizes Autonomous Shipping. *Maritime Executive*, 6 September 2018. https://maritime-executive.com/article/iacs-chairman-prioritizes-autonomous-shipping#gs.ReflfK0.
MAREX 2018. HNA Group Forms Chinese Alliance for Autonomous Ships. *Maritime Executive*, 28 July 2017. https://maritime-executive.com/article/hna-group-forms-chinese-alliance-for-autonomous-ships #gs.CetnW_c.
MAREX 2019. Autonomous Collision Avoidance Tested Using Damen Crew Boat. *Maritime Executive*, 26 March 2019. https://www.maritime-executive.com/article/autonomous-collision-avoidance-tested-using-damen-crew-boat.
Maritime UK 2017. Being a Responsible Industry. An Industry Code of Practice, Maritime Autonomous Surface Ships up to 24 metres in Length. A Voluntary Code, Version 1.0, November 2017.
Maritime UK 2018. Call for Evidence. Department for Transport, London, 27 March 2018. www.gov.uk/government/consultations/maritime-2050-call-for-evidence.
MLIT 2018. Japan's MLIT Selects MOL's Autonomous Shipping Project. Ship Technology, 7 August 2018. https://www.ship-technology.com/news/japans-mlit-selects-mols-autonomous-shipping-project/.
MPA 2018. MPA Expands MOU with DNV GL to include Autonomous Systems and Intelligent Shipping. Maritime Port Authority of Singapore. Media Release, 3 February 2017. https://www.mpa.gov.sg/web/portal/home/media-centre/news-releases/detail/6172e16c-583f-4b79-8fa9-d7f3adca81ca.
MPA 2018-1. Singapore to Develop Autonomous Vessels. *World Maritime News*, 25 April 2018. https://worldmaritimenews.com/archives/251143/singapore-to-develop-autonomous-vessels/.
MPA 2018-2. MPA announces record seven agreements at Singapore Maritime Technology Conference. Maritime Port Authority of Singapore. *News Release*, 25 April 2018. https://www.mpa.gov.sg/web/portal/home/media-centre/news-releases/detail/dc2beca8-df1b-427f-843a-b1160148a808.
MSC 98/20/2. Work Programme. Maritime Autonomous Surface Ships Proposal for a Regulatory Scoping Exercise. Regulatory Scoping Exercise for the Use of Maritime Autonomous Surface Ships (MASS). 27 February 2017.
MSC 99/5. Comments on the Regulatory Scoping Exercise. Secretariat.
MSC 99/5/1. Comments and Proposals on the Way Forward for the Regulatory Scoping Exercise. IFSMA and ITF, 22 February 2018.
MSC 99/5/2. Proposals for the Development of a Work Plan. International Chamber of Shipping, 8 March 2018.

MSC 99/5/3. Recommendations on Identification of Potential Amendments to Existing IMO Instruments. Finland, Liberia, Singapore, South Africa and Sweden, 8 March 2018.
MSC 99/5/4. Conosiderationos on and Proposals for the Methodology to use within the Framework of the Regulatory Scoiping Exercise. France. 28 March 2018.
MSC 99/5/5. Plan of Approach for the Scoping Exercise. Australia et al., 12 March 2018.
MSC 99/5/6. Considerations on Definitions for Levels and Concepts of Autonomy, Finland.
MSC 99/5/7. Proposal on the Work Plan of the Regulatory Scoping Exercise for the Use of MASS. China and Finland, 13 March 2018.
MSC 99/5/8. Recommendations on Categorization and Regulatory Scoping Exercise of MASS. China and Liberia, 13 March 2018.
MSC 99/5/9. Japan's Perspective on Regulatory Scoping Exercise for the Use of MASS. Japan, 13 March 2018.
MSC 99/5/10. General Comments on a Way Forward. ITF, 28 March 2018.
MSC 99/5/11. Regulatory Scoping Exercise for the Use of Maritime Autonomous Surface Ships (MASS).
MSC 99/5/12. Comments on Documents MSC 99/5, MSC 99/5/2, MSC 99/5/5, MSC 99/5/8 and MSC 99/5/9. Turkey, 27 March 2018.
MSC 99–22. Report of the Maritime Safety Committee on Its Ninety-Ninth Session, 5 June 2018.
MSC 99/22a. Report of the Maritime Safety Committee on Its Ninety-Ninth Session, p. 36.
MSC 99/INF.3. Final Report: Analysis of Regulatory Barriers to the Use of Autonomous Ships. Denmark.
MSC 99/INF.13. Establishing International Test Area "Jaakonmeri" for Autonomous Vessels, Finland. 12 March 2018
MSC 99/INF.14. Studies Conducted in Japan on Mandatory Regulations Relating to Maritime Autonomous Surface Ships – SOLAS, STCW and COLREGs. Japan, 13 March 2018.
MSC 99/INF.16. Presentation by Norway on 21 May 2018 on the "YARA Birkeland" Development. Norway, 13 March 2018.
MSC 100.Sum. www.imo.org/en/MediaCentre/MeetingSummaries/MSC/Pages/MSC-100th-session.aspx.
MSC 100/5/1. Proposal for a Classification Scheme for Degrees of Autonomy. International Standards Organization, 31 August 2018.
MSC 100/5/2. Interim Guidelines for MASS Trials. Norway and BIMCO, 28 September 2018.
MSC 100/5/3. Proposals for the Development of Interim Guidelines for Maritime Autonomous Surface Ships (MASS) Trials. Republic of Korea, 28 September 2018.
MSC 100/5/4. Comments on Document MSC 100/5. IMO Secretariat, 12 October 2018.
MSC 100/5/5. Comments on Document MSC 100/5. Japan, 12 October 2018.
MSC 100/5/6. Comments on Document MSC 100/5. Australia, Denmark, Finland, France and Turkey, 12 October 2018.
MSC 100/5/7. Comments on Document MSC 100/5. China, 12 October 2018.
MSC 100/5/8. Comments on Document MSC 100/5. United States, 12 October 2018.

MSC 100-INF.3. Initial Review of IMO Instruments under the Purview of MSC. IMO Secretariat.
MSC 100-INF.6. Preliminary Analysis of the International Regulations for Preventing Collisions at Sea, China.
MSC 100-INF.10. Results of Technology Assessment on MASS. Korea, 28 September 2018.
MSC 101.Sum. www.imo.org/en/MediaCentre/MeetingSummaries/MSC/Pages/MSC-101st-session.aspx.
MSC 101/5/1. Comments and Proposals for Interim Guidelines for MASS Trials. ITF, 28 February 2019.
MSC 101/5/5. Interim Guidelines for MASS Trials. Finland, Japan, Norway, Republic of Korea, Singapore, United Arab Emirates and BIMCO, 2 April 2019.
MSC 101/WP.8. Report of the Working Group (draft). Regulatory Scoping Exercise for the Use of Maritime Autonomous Surface Ships (MASS). Maritime Safety Committee. 101st session. Agenda Item 5, 12 June 2019.
MUNIN 2016. Final Brochure. The Project Maritime Unmanned Navigation through Intelligence in Networks (MUNIN), 2016. http://www.unmanned-ship.org/munin/.
NI 2017. Autonomous Ships and the Human Element. *Maritime Executive*, 11 September 2017. https://maritime-executive.com/corporate/autonomous-ships-and-the-human-element#gs.=cuZODk.
NI 2018. Autonomous Ships should Comply with COLREGs Says The Nautical Institute. Press Release, 7 June 2018. https://www.nautinst.org/en/about-the-institute/press/index.cfm/autocolregs.
*Nikkei Asian Review* 2017. Japan Aims to Launch Self-piloting Ships by 2025. 8 June 2017. https://asia.nikkei.com/Tech-Science/Tech/Japan-aims-to-launch-self-piloting-ships-by-2025.
Norwegian Forum for Autonomous Ships (NFAS) 2018. www.nfas.autonomous-ship.org.
One Sea 2018. One Sea Autonomous Marine Ecosystem. www.oneseaecosystem.net.
Organization Internationale de Normalization (ISO), 2018. https://www.iso.org/about-us.html.
Ship Technology 2018. China to Build Unmanned Ship Test Facility in Zhuhai. 13 February 2018. https://www.ship-technology.com/news/china-build-unmanned-ship-test-facility-zhuhai/.
SmartShips 2018. https://smartshipscoalition.org.
SmartShips 2018–1. Marine Autonomy Research Site (MARS). SmartShips Coalition. https://smartshipscoalition.org/maritime-autonomy-research-site-mars/.
Tekes 2016. Press Release by the Finnish Funding Agency for Technology and Innovation (Tekes), 23 September 2016: Autonomous maritime ecosystem starts in Finland.
White, Travis 2018. Marine Autonomy Research Site Inaugural Experiment Answers Questions for DARPA to Advance Autonomous Surface Vessel Technology. Great Lakes Research Center. Smart Ships Coalition, October 2018. https://smartshipscoalition.org/2018/10/26/marine-autonomy-research-site-mars-inaugural-experiment-answers-questions-for-defense-advanced-research-projects-agency-darpa-to-advance-autonomous-surface-vessel-technology/.

Wingrove, M. 2018. Autonomous Vessels will Revolutionise Shipping. Maritime Digitalisation & Communications, 8 March 2018. http://www.marinemec.com/news/view,autonomous-vessels-will-revolutionise-shipping_51036.htm.

World Maritime News 2017. Unmanned Cargo Ship Development Alliance Launched in Shanghai, 29 June 2017. https://worldmaritimenews.com/archives/223982/unmanned-cargo-ship-development-alliance-launched-in-shanghai/.

# Chapter 11
# Legal Issues

The regulatory process of the International Maritime Organization (IMO) and member nation states described in the previous chapter results in a set of legal instruments in the form of conventions to adopt policy that, when accepted by governments, becomes national law and is subject to enforcement. There are many international conventions and many additional codes, protocols, agreements and other instruments promulgated by the IMO that are presently adopted and in force worldwide [IMO CONV]. Some of these are under the purview of the Maritime Safety Committee (MSC) while others fall under the committees concerned with security and piracy, legal affairs and the environment. These include the three key IMO conventions related to Safety of Life at Sea (SOLAS), Prevention of Pollution from Ships (MARPOL) and Standards of Training, Certification and Watchkeeping for Seafarers (STCW) along with conventions relating to maritime safety and security, the ship/port interface, prevention of pollution, liability and compensation and other topics.

The regulatory scoping exercise for the use of Maritime Autonomous Surface Ships (MASS) established in June 2017 at the 98th session of the Maritime Safety Committee (MSC 98) instituted a two-step process [LEG 106/8/1]. The first step was designed to survey these legal instruments and identify provisions with them which, as currently drafted:

(A) apply to MASS and prevent MASS operations; or
(B) apply to MASS and do not prevent MASS operations and require no actions; or
(C) apply to MASS and do not prevent MASS operations but may need to be amended or clarified, and/or may contain gaps; or
(D) have no application to MASS operations.

Once the first step is completed, a second step is conducted to analyze and determine the most appropriate way of addressing MASS operations taking into account, among other things, the human element, technology and operational factors by:

- equivalences as provided for by the instruments or developing interpretations; and/or
- amending existing instruments; and/or
- developing new instruments; or
- none of the above as a result of the analysis.

The first step of this regulatory scoping exercise continued with the development and consideration of comments by volunteering member states that was completed in September 2019 at the Intercessional MSC Working Group meeting with the preliminary results of this exercise described in this chapter. The second step focused on analyzing the most appropriate way to address MASS operations followed by a commenting stage, with the presentation of results and final consideration occurring at the 102nd meeting of the MSC (MSC 102) in May 2020 [LEG 106/8/1, p. 9].

Recommendations have been provided for instruments that are under the purview of the Maritime Safety and other IMO committees. This chapter attempts to summarize many of the topics and comments on instruments relevant to the operation of MASS and also identifies instruments that appear to have little or no application to MASS operations. The list of instruments and the volunteering members undertaking and supporting the review of those instruments are identified based upon the MSC 101 Status Report of the Regulatory Scoping Process and the review results cited [MSC 101/5].

## 11.1 INSTRUMENTS REQUIRING AMENDMENTS TO SUPPORT MASS OPERATIONS

During the regulatory scoping exercise focus placed on the review of SOLAS, with participating member states performing initial reviews on a chapter-by-chapter basis covering many of the conventions, codes, protocols and agreements emanating from SOLAS. The comments in this paragraph reflect those initial reviews as well as comments pertaining to these reviews received during the comment period. Comments that are general in nature and that may apply across several chapters and codes such as e-certificate policy and the need for trained seafarers to operate life rafts are listed as they first appear and are usually not repeated. This list is not exhaustive and several conventions and codes were not reviewed but are listed for the sake of completeness.

Common assumptions made related to the regulatory scoping exercise include for IMO Degrees One and Two Autonomy that seafarers are on board and assumed to be available to perform the various human intervention actions that are currently built into the regulations for emergency and casualty situations that may include manual operation and control of valves monitoring and responding to alarms and other. For IMO Degrees Three and Four Autonomy it is assumed that persons other than seafarers may stay on board including passengers, special personnel or technical personnel temporarily embarked for the purpose of maintenance or time-limited technical intervention.

### 11.1.1 International Convention for the Safety of Life at Sea (SOLAS Convention)

The International Convention for the Safety of Life at Sea (SOLAS) 1974, as amended, is generally viewed as the most important of all international treaties concerning the safety of merchant ships [SOLAS 1974]. The first version was adopted in 1914 in response to the *Titanic* disaster and has been updated and amended on numerous occasions. Consisting of 14 chapters, its purpose is to specify minimum standards for the construction, equipment and operation of ships compatible with safety. Several of these chapters have spawned many of the conventions and codes described. A summary of review comments is provided in the paragraphs that follow.

*Chapter I General Provisions* – The survey of ships, issuance of documents signifying the ship meets the requirements of the Convention and the control of ships in ports are described. This chapter was not reviewed as part of the regulatory scoping exercise.

*Chapter II-1 Construction: Subdivision and Stability, Machinery and Electrical Installations* – Topics covered in this chapter relate to the subdivision of ships into watertight compartments, the maintenance of machinery and electrical installations, and construction for a specified design life to minimize the loss of ship and to be environmentally friendly in the event of structural failure. The initial review of this chapter was performed by France with support from China, Iran and Sweden [ISWG/MASS 1/2/1].

The results of the initial review indicate of a total of 324 records:

(B) 196 (61%) apply to MASS, do not prevent MASS operations and require no actions.
(C) 108 (33%) apply to MASS and do not prevent MASS operations but may need to be amended or clarified, and/or may contain gaps.
(A) 10 (3%) apply to MASS and prevent MASS operations, and
(D) 10 (3%) have no application to MASS operations.

Under IMO Degree One Autonomy almost all regulations were identified as (B) apply to MASS, do not prevent MASS operations and require no actions. General comments were made recommending amendment or regulation in a separate instrument MASS control and operation including presentation of indications and alarms can be located not on board the ship, and several tasks normally managed manually by the master/crew can be fully automated.

For IMO Degrees Three and Four Autonomy most of the regulations that were identified as (C) applicable to MASS and do not prevent MASS operations but may need to be amended or clarified, and/or may contain gaps may require amendment or clarification as these regulations require manual operations or indication/alarm on the bridge. Specific definitions should also be added for MASS operations.

*Chapter II-2 Construction: Fire Protection, Fire Detection and Fire Extinction* – Detailed fire safety provisions for all ships and specific measures for passenger ships, cargo ships and tankers are contained in this chapter. The initial review was submitted by Japan with support by China and the International Association of Classification Societies Ltd (IACS) [ISWG/MASS 1/2/3; ISWG/MASS 1/2/6].

Comments included a general observation that the meanings of "master", "crew", "responsible person", etc., should be clarified for IMO Degrees of Autonomy Two, Three and Four, taking into account the possibility that personnel from the ship are not on board. They also indicated that for IMO Degrees of Autonomy Two, Three and Four, the definitions of "Control Stations" and "Safety Center" should be amended to introduce remote control center or remote location for supervision. Also, the terminology "Safety center on passenger ships" should be amended since the safety center could be remote and the provision for the functionality of the safety systems required to be available from the Safety center should be amended to include automated or autonomous systems.

For Degrees of Autonomy Two, Three and Four, the definitions of manned spaces should be amended and since the decision making will be done remotely, either autonomously or automatically, additional functional requirements may be needed to demonstrate that the remote control center or the autonomous or automated system can detect and control fire.

For Degrees of Autonomy Three and Four many provisions requiring manual operations and other actions by personnel on board, e.g., firefighting, and some provisions regarding accommodations, accessibility, alarms and safety centers were identified as applying to MASS and do not prevent MASS operations but may need to be amended or contain gaps. A future issue to be addressed includes how to evaluate the reduction of risks due to the absence of persons on board. A comment was made that unmanned passenger transports cannot be put into practice since life rafts would require certified life raft crew and personnel to assist in evacuation and possibly firefighting. For this reason MASS without seafarers should not be considered for operation as passenger ships.

Other comments included that under Degrees Three and Four Autonomy catastrophic consequences would result, especially for cargo ships, in the event of fire cutting off communication with the remote control center and whether the remote location could be the continuously manned central control station. Also considered was whether the designated stations can shift between the remote location and a ship. Regulations may also have to amended, clarified or contain gaps pertaining to smoke generation potential and toxicity, detection and alarm, control of smoke spread, containment of fire and notification of crew and passengers. Guidelines related to operational readiness and maintenance may also have to be reviewed [MSC.1/Circ. 1432]. Existing instructions pertaining to onboard training and drills may prevent MASS operations and may need to be redrafted entirely or amended to designate responsible parties for fire drills other than seafarers when carrying passengers or other personnel. Other comments pertain to helicopter facilities that require firefighting personnel, which may prevent MASS operations, the requirement for fire patrol that requires crew and the protection provided by portable extinguishers and manual firefighting equipment. There are also inconsistencies and gaps pertaining to the casualty threshold for safe return to port and safe areas in addition to requirements for fire patrols in passenger ships and an efficient fire patrol system for special category spaces.

*Chapter III Life-saving Appliances and Arrangements* – This chapter covers requirements for life boats, rescue boats and life jackets according to the type of ship. Specific technical requirements for life-saving appliances are given in the Life-Saving Appliance (LSA Code). The initial review was submitted by Belgium and the Netherlands with support by China [ISWG/MASS 1/2/11].

The results of the initial review indicated that the biggest challenge is when passengers are transported on a passenger ship or a cargo ship (up to 12 passengers), seafarers are required on board to assist and/or evacuate those passengers in an emergency. Unless future means could be developed to provide such functions in an automated manner, this inhibits the operation of MASS under Degrees Three and Four Autonomy as the on board presence of certificated persons is necessary to perform those functions.

*Chapter IV Radio Communications* – This chapter describes requirements for radio equipment including Global Maritime Distress Safety System (GMDSS), satellite Emergency Position Indicating Radio Beacons (EPIRB) (EPIRBs) and Search and Rescue Transponders (SARTs). The review of this chapter was submitted by Turkey with support by China and Japan [ISWG/MASS 1/2/15].

The findings of the initial review indicated that the vast majority of all regulations in this chapter apply to all Degrees of Autonomy. Depending on the Degree of Autonomy it was agreed that some of the regulations do not prevent MASS operations, some contain gaps or need clarification and only few are likely to prevent MASS operations.

Regulations on radio communications will have significant importance for MASS operations and additional findings resulting from this exercise are listed including a need to consider functional and maintenance requirements once they are introduced within the concept of MASS, technical requirements for remote control stations and main control stations, radio watch requirements and radio personnel for IMO Degrees Three and Four

Autonomy, and distress, safety and urgency calls and related requirements taking into account the technological developments.

*Chapter V Safety of Navigation* – This chapter considers requirements that all vessels are sufficiently and efficiently manned from a safety perspective. The review of this chapter was submitted by China and Singapore with support from Denmark and Japan [ISWG/MASS 1/2/16].

The results of the initial review determined that almost all of the regulations in this chapter were found to be applicable to MASS. The majority of them do not prevent the operation of MASS and have either been identified "as apply to MASS and require no actions" or "apply to MASS but may need to be amended".

Additional comments note that there are no requirements for remote control in current regulations. Since remote control is essential for the operation of MASS degrees two and three, new regulations are required, especially the requirements on function, design, electromagnetic compatibility, visibility, manning, training and drilling, and information transmission, etc. Requirements on manning, responsibilities, qualification, and necessity of the ship Master for MASS for each Degree of Autonomy should be amended or clarified, especially for remotely controlled and fully autonomous ships. Current definitions in this chapter are considered applicable. However, gaps might exist for MASS operation and therefore review, and amendments are needed, such as the introduction of a MASS definition as well as other possible new definitions due to the amendment of regulations in this chapter.

Relevant findings of the comments have been identified for regulations regarding administration such as services and warnings and, since they are closely related to MASS operation and essential for MASS navigational safety, they have been considered as applicable to MASS during initial review. Nonetheless, there were comments deeming that they do not apply to MASS operation as these requirements for services and warnings apply to governments not ships.

*Chapter VI Carriage of Cargoes* – All types of cargo except for liquids and gasses in bulk, including requirements for stowage and securing these cargo, are discussed in this chapter. The review of this chapter was submitted by Japan with support by China [ISWG/MASS 1/2/4].

In the initial review provisions were identified as (C) applying to MASS and do not prevent MASS operations but may need to be amended or clarified, and/or may contain gaps. This includes some provisions under Degrees of Autonomy Three and Four requiring manual operations and other actions by personnel on board, e.g., emergency response and onboard inspection, where appropriate alternative safety measures should be adopted to achieve the functionalities intended by the existing regulations. However, to ensure safety a future issue to be addressed is how to evaluate the reduction of risks owing to absence of persons on board. Some provisions require action by personnel. For the carriage of cargoes by ships without persons on board during sailing, one of the big challenges is how to establish emergency procedures to deal with conditions of leakage, spillage or fire involving cargoes. Another future issue to be considered is the introduction of different procedures when no persons are on board and the cargo does not include any substances harmful to the marine environment.

*Chapter VII Carriage of Dangerous Goods* – The parts of this chapter cover the carriage of dangerous goods and the construction of ships for this purpose. The review of this chapter was submitted by Japan with support by China [ISWG/MASS 1/2/5].

The results of the initial review indicate regulations 1, 3, 7, 7-5 and 8 through 16, (D) appear to have no application to MASS operations. Regulations 2, 4, 5, 7-1, 7-2, and 7-3 were identified as (C) applying to MASS and do not prevent MASS operations but may need to be amended or clarified, and/or may contain gaps for Degrees of Autonomy Three and Four. This includes provisions requiring manual operations, e.g., starting of pumps supplying the water-spray system, and some provisions regarding accommodations and accessibility. Also included are some provisions requiring actions by personnel on board, e.g., emergency response and onboard inspection. Clarifications are needed related to emergency response and medical first aid, documents required for carriage of dangerous goods, fully autonomous loading and unloading, and the monitoring of containers and vehicle lashings over the course of voyages. A gap appears to exist as to whether detailed instructions on emergency response and medical first aid relative to incidents involving dangerous goods in solid form in bulk were necessary. Additional gaps may exist as to the transport document, special list, manifest, or stowage plan being retained on board is necessary for an autonomous vessel, and how MASS would perform the required task of monitoring stowage and segregation that was typically the responsibility of the master of the vessel. Regulations 6 and 7-4 appear to have a gap as to whether MASS could determine the loss of dangerous goods and report the particulars of the loss as required.

*Chapter VIII Nuclear ships* – Requirements for nuclear powered ships to conform to the regulations proscribed in the Code of Safety for Nuclear Merchant Ships are described. This chapter and the corresponding code were not reviewed as part of the regulatory scoping exercise for MASS.

*Chapter IX Management for the Safe Operation of Ships* – This chapter requires ship owners, persons and companies that have assumed the responsibility for a ship to comply with the International Safety Management (ISM Code). The review of this chapter was submitted by Norway with support by China, Nigeria, the Republic of Korea and the Russian Federation [ISWG/MASS 1/2].

The results of the review indicate that the instruments in the context of requirements adopted in chapter IX and the associated ISM Code there are no critical or substantial regulatory barriers which prevent the safe operation of MASS. Requirements adopted in relevant regulations of SOLAS and associated instruments presuppose ships are operated by an onboard crew complement. Although at present certain activities are performed on board to attain the functions required by the relevant instruments, such activities may be carried out remotely from the ship, which then will fulfill relevant prescriptive or functional requirements.

Requirements associated with the safety of persons on board a ship may be considered to apply only when such persons are present on board. Conversely, such requirements will naturally not apply in the absence of such persons, such as for a category three or four MASS where no persons are intended to be on board. A recommendation was made that the development and the subsequent adoption of a mandatory Code for MASS should be considered as a road map going forward.

*Chapter X Safety Measures for High-Speed Craft* – This chapter mandates the International Code of Safety for High-Speed Craft (HSC), which were not reviewed as part of the regulatory scoping exercise for MASS.

*Chapter XI-1 Special Measures to Enhance Maritime Safety* – This chapter covers requirements for carrying out surveys, inspections, ship numbering and other requirements. The initial review was submitted by Finland with support from China [ISWG/MASS 1/2/12].

The results of the initial review considered that passenger transports without seafarers on board cannot be performed as life rafts would require certified life raft crew and personnel to assist in evacuation and possibly firefighting. The requirements of this chapter do not pose any specific design, construction, equipment or operational requirements and can be applied to all MASS without amendments.

*Chapter XI-2 Special Measures to Enhance Maritime Security* – This chapter covers requirements for ship security including the International Ship and Port Facility Security Code (ISPS) Code. The initial review was submitted by Finland with support from China [ISWG/MASS 1/2/13].

The results of this review are based upon an assumption that regulations in this chapter pertaining to alternative and equivalent security agreements would allow for creating special shipboard plans and arrangements for unmanned ships. This would help ensure the integrity this chapter and the related Code regarding its applicability to traditionally manned ships, and not create confusion on how the initial requirements are to be applied. This topic will need further discussion. It was noted there might be a need to add the term "autonomous ships" to the definitions.

Other comments considered that the physical absence of the master on board would prevent MASS operations, and that port state control procedures and actions were seen to apply but need to be considered from a procedural aspect when there are no persons available on board while intending to enter a port of another contracting government.

*Chapter XII Additional Safety Measures for Bulk Carriers* – Topics covered in this chapter include standards for bulk carrier bulkhead and double bottom strength, owners' inspection and maintenance of bulk carrier hatch covers, and criteria for side structures of bulk carriers of single-side skin construction. The review of this chapter was submitted by Japan [ISWG/MASS 1/2/7].

The results of the initial review included regulations 3, "Implementation schedule", and 9, "Requirements", for bulk carriers not being capable of complying with regulation 4.3 due to the design configuration of their cargo holds were identified as (D) having no application to MASS operations, regardless of Degree of Autonomy since they only apply to existing ships constructed on or before specific dates. All other regulations were identified as (B) applying to MASS and do not prevent MASS operations and require no actions.

*Chapter XIII Verification of Compliance* – This chapter mandates implementation of the IMO Member State Audit Scheme. The review of this chapter, excluding provisions related to the III Code, was submitted by Japan [ISWG/MASS 1/2/8].

The initial review determined all regulations were identified as (B) applying to MASS and do not prevent MASS operations and require no actions regardless of Degree of Autonomy. Comments received during the comment period indicated a belief that the regulations have no application to MASS operations but the results of the initial review remain unchanged.

*Chapter XIV Safety Measures for Ships Operating in Polar Waters* – This chapter mandates the introduction and part I-A of the International Code for Ships Operating in Polar Waters (Polar Code). The review of this chapter was submitted by Finland [ISWG/MASS 1/2/14].

This chapter (B) applies to MASS and requires no actions for all Degrees of Autonomy. A comment was raised that clarification was needed regarding recording alternative design and arrangements in the Polar Water Operational Manual. However, this provision appears sufficient in its current form without amendments since the requirement is general and can be performed by any person having sufficient knowledge of the vessel and circumstances concerned.

## 11.1.2 International Ship and Port Facility Security (ISPS Code)

This review covers requirements for ship security including the International Ship and Port Facility Security Code (ISPS Code). The initial review was submitted by Finland with support from China [ISWG/MASS 1/2/13].

The results of this review indicated that the ISPS Code (B) applies to MASS and requires no amendments because the recognized gaps can be solved taking into account SOLAS regulations XI-2/11 (Alternative security agreements) and XI-2 12 (Equivalent security arrangements). Comments included that under IMO Degree One Autonomy ISPS Code Section A/2 (Definitions) might need to be amended or clarified and/

or may contain gaps and that the Ship Security Plan required under the Code could establish the relevant persons and obligations for the functions required when no persons are on board. Obligations are not directly established by the Code to be based on board rather than by defined persons. Also considered was that the ship might be protected and secured by a dedicated a electronic surveillance, protection and alarm system, as any industrial and/or financially important facility on shore and would not be relying on human presence. Such a system could be remotely supervised and controlled. Functions, rights and responsibilities of remote operating centers should be defined in the ship security plan.

### 11.1.3 International Safety Management (ISM Code)

The International Safety Management (ISM) Code provides an international standard for the safe management and operation of ships and for pollution prevention. The review of this Code was submitted by Norway with the support of China, Nigeria, the Republic of Korea and the Russian Federation [ISWG/MASS 1/2].

The initial review of the Code determined the instruments may be applied for the safe operation of MASS. There were no significant differences between the initial review and comments received during the comment period. A recommendation was made for the development and the subsequent adoption of a mandatory Code for MASS should be considered as a road map going forward.

### 11.1.4 International Convention on Standards of Training, Certification and Watchkeeping for Seafarers (STCW Convention)

The International Convention on Standards of Training, Certification and Watchkeeping for Seafarers (STCW) and Code, 1978, as amended including the 1995 and 2010 Manila Amendments, provides sets minimum qualifications and regulations for masters, officers and watch personnel on seagoing merchant ships and large yachts. The review of the STCW was submitted by United States with support by China, Cyprus, Japan, the Republic of Korea, and Russian Federation and Spain [ISWG/MASS 1/2/20].

The exercise examined the STCW Convention and Code concurrently since the Convention specifies the requirements and Code provides the minimum standard to be maintained by the Parties to give full and complete effect to the Convention requirements. The term "seafarer" with respect to the Degrees of Autonomy is understood to mean those persons trained and qualified to perform vessel operational duties and responsibilities in accordance with the STCW Convention. The term "remote operator" for the purpose of the regulatory scoping exercise is understood to mean a person not located on board the ship for which the training and qualifications of this person are not provided.

Assumptions for IMO Degrees of Autonomy Three and Four are dependent on duties and responsibilities derived from the carriage and operational requirements in other Conventions. For example, STCW will apply when seafarers such as AB-deck and engine may be aboard a Degree Three ship to conduct maintenance work while on a Degree Four ship there may be security personnel aboard. The regulatory scoping exercise does not take into account these options. The general assumption is that the "remote operator" operates and controls shipboard systems and functions. The initial review found that the Convention and Code requirements remain valid when there are seafarers on board and do not apply when there are no seafarers on board.

### 11.1.5 Convention on the International Regulations for Preventing Collisions at Sea (COLREG)

The Convention on the International Regulations for Preventing Collisions at Sea (COLREG) 1972 establish navigation rules to be followed by ships and other vessels to prevent collisions between two or more vessels. The review of this Convention was submitted by Marshall Islands with support from China, Japan, Singapore, Spain, Sweden, the United States and the non-governmental organization (NGO) GlobalMET [ISWG/MASS 1/2/19].

The results of the initial review pertaining to Degree One Autonomy cite the MASS definition where "Seafarers are on board to operate and control shipboard systems and functions. Some operations may be automated and at times unsupervised but with seafarers on board ready to take control" as conflicting with manning requirements provided in SOLAS regulation V/14 and the STCW Convention ) Chapter VIII, Part 4.1, where it is stated, "at times the vessel will be unsupervised". This statement related to bridge watchkeeping illustrating a critical difference between current conventional ship's in that the automated systems on board may potentially be solely responsible for navigation for some periods on the bridge and whether Degree One MASS would comply with the COLREG as they currently stand. Rule 5 would another example of this where it states: "every vessel shall at all times maintain a proper look-out by sight and hearing". The question is whether automated systems can be considered a proper look-out. It is not expressly stated within the COLREG that this is the case. However, this Convention did not perhaps foresee this circumstance when it was established, and human centric principles are at the heart of this convention.

Under Degree Two Autonomy a main issue interpreted differently by the group was the role of the seafarers onboard MASS when the ship is controlled and operated from another location. Of specific concern was the point at which seafarers will be able to take control and operate shipboard systems and functions and how will this process be regulated. Another issue included potential problems related to signal interpretation and response based on the possibility of remote control and operation when trying to interpret and transmit audible signals when seafarers will be present on board but not on the bridge at all times. As a result, the transfer of information to a remote control center will be vital in order to comply with the COLREG. It was noted that a possible way to deal with this issue could be the use of Automated Identification System (AIS) information exchange to replace the required sound signals.

Under Degrees Three and Four Autonomy where the ship is controlled and operated from another location and there are no seafarers on board, this represents a large change for the industry and as a result a number of potential issues requiring clarification were observed. This included the possibility that an unmanned MASS will be constructed differently from a conventional ship requiring a separate section within the annexes similarly to Annex I/13 with high-speed craft. Annex IV also raised significant discussion regarding the ability for an unmanned MASS to indicate distress that will require further clarification and discussion. Questions exist requiring clarification and raising concern as to whether a remote-controlled operator can assume the role of "Master or crew", and regarding the ability of a remote-controlled operator to reach the same standards required of them as a look-out onboard particularly during difficult weather conditions and sea states and the ability to detect smaller vessels that perhaps radar would struggle to identify. Another issue relates to the need for a constant connection between the remote operator and the vessel itself where disturbances or losses of connection may directly prevent the remote operator from maintaining "a proper look-out". This also applies to Rule 19 (Conduct of vessels in restricted visibility) and further clarification will be necessary.

IMO Degree Four Autonomy, where the operating system of the ship is able to make decisions and determine actions by itself, raised a further question requiring clarification as to a need for the "electronic look out" to be able to detect and understand the established sound and light signals from other vessels and act accordingly. It cannot be underestimated the obligation for this technology to reach (at least) the same look-out standards. It was discussed that this rule could prevent MASS operations if it could not be concluded that a fully autonomous vessel could maintain a proper look-out by sight and hearing; however, the group agreed that further discussion was necessary. The group felt that the nature of Rule 6 (Safe speed) and its requirements were not intended to be applied to fully autonomous vessels and as a result this rule contains a gap. Rule 8 ("Action to avoid collision") also raises a similar problem regarding whether a fully autonomous vessel can apply "good seamanship" principles when navigating and further discussion is needed on what clarification or amendment may be required. Rule 18 ("Responsibilities between vessels") by its nature requires understanding of the types of vessels involved and questions were raised regarding the ability of a fully autonomous vessel to identify these different vessel types and act accordingly. Rule 19 ("Conduct of vessels in restricted visibility") displayed the complexities for fully autonomous ships to be able to comprehend, interpret and comply through the application of multiple rules requiring further discussion. Additional comments were made regarding terminology; lights, shapes and sound signals; COLREG compliance; and a sailing vessel could be a MASS and its interactions with other vessels and other MASS.

### 11.1.6 International Convention on Maritime Search and Rescue (SAR Convention)

The International Convention on Maritime Search and Rescue (SAR Convention) 1979 developed an international SAR plan that, regardless of where an accident occurs, the rescue of persons in distress at sea may be coordinated by a SAR organization. The review of the SAR Convention was submitted by France and Spain and supported by Turkey [ISWG/MASS 1/2/2].

For the purpose of the regulatory scoping exercise, MASS were not considered as a potential search and rescue unit. Such development should be examined specifically. However, as any vessel, a MASS could be involved in SAR operations in application of the general obligation to assist persons in distress.

For chapters 1, 4 and 5, IMO Degree of Autonomy One was considered as (B) applying to MASS and do not prevent MASS operations and require no actions. IMO Degrees Two, Three and Four Autonomy were considered (C) applying to MASS and do not prevent MASS operations but may need to be amended or clarified, and/or may contain gaps. For chapters 2 and 3 Degrees of Autonomy One and Two were considered as (B) applying to MASS and do not prevent MASS operations and require no actions. Degrees Three and Four Autonomy were considered (C) applying to MASS and do not prevent MASS operations but may need to be amended or clarified, and/or may contain gaps.

## 11.2 SOME SIGNIFICANT CHANGES TO SUPPORT MASS OPERATIONS

The following instruments are not expected to have major significance in the implementation of MASS but should be considered with respect to definitions, commonality of terms and other issues in common between MASS and conventional ships.

### 11.2.1 International Code for Ships Operating in Polar Waters (Polar Code)

The International Code for Ships Operating in Polar Waters (Polar Code) provides additional regulations for safety acknowledging that polar waters may impose additional demands on ships beyond those normally encountered. The review of the Polar Code was submitted by Finland [ISWG/MASS 1/2/14].

The interim review results indicate that the Polar Code applies to MASS and requires no actions for IMO Degrees of Autonomy One and Two. The requirements in the Polar Code regarding onboard certificates and life-saving appliances should be clarified for MASS operating under IMO Degrees Three and Four. One comment was received indicating that there were no seafarers on board, the provisions for life-saving appliances and voyage planning of the Polar Code would need to be amended. However, the Polar Code is considered to apply without amendments since it is an add-on to the requirements specified in the SOLAS Convention and these requirements apply only if the equipment is fitted.

### 11.2.2 International Maritime Dangerous Goods (IMDG Code)

The IMDG Code is a two-volume code comprising parts A and A-1 of Chapter VII of the 1974 SOLAS Convention where compliance is required for the transport of dangerous goods by ship. The review comments were submitted by Japan with support by China [ISWG/MASS 1/2/6a].

Comments include a need to amend Part 7, "Provisions concerning transport operations", which requires regular inspections of ro-ro cargo spaces during the voyage by an authorized crew member or responsible person for early detection of any hazard and refers to the judgment by the master in the event of incidents. Also, requirements for "on deck stowage" and in Part 3, "Dangerous goods list, special provisions and exceptions", should be amended as constant supervision and accessibility will not be maintained, or if there is a substantial risk of formation of explosive gas mixtures, development of highly toxic vapors, or unobserved corrosion of the ship on the ships of Degrees of Autonomy Three and Four.

### 11.2.3 International Bulk Chemical (IBC Code)

The International Bulk Chemical (IBC) Code addresses the design, construction and outfitting of newly built or concerted chemical tankers. The review of the IBC was performed as part of the review of SOLAS Chapter II-2 submitted by Japan [ISWG/MASS 1/2/6].

Comments resulting from the review include in IBC chapter 11, "Fire protection and fire extinction", the use of foam applicators, fire mains and hydrants, and dry chemical powder fire-extinguishing systems deemed unusable during sailing when there are no seafarers on board. Also, a water fire-extinguishing system and fixed fire-extinguishing arrangements for category A machinery spaces of category may not be effective when there are no seafarers on board. Additional comments include in chapter 15 citing potential for amendment where a requirement exists for local manual operation of the control system of any cooling system, remote manual operation for remote starting of pumps supplying the water-spray system, and remote operation of any normally closed valves in the system from a location adjacent to the accommodation spaces. Also considered for amendment is chapter 16, "Operational requirements", which requires officers be trained in emergency procedures to deal with conditions of leakage, spillage or fire involving the cargo and a sufficient number of them shall be instructed and trained in essential first aid for cargoes carried.

### 11.2.4 International Code for the Construction and Equipment of Ships Carrying Liquefied Gasses in Bulk (IGC Code)

The International Code for the Construction and Equipment of Ships Carrying Liquefied Gasses in Bulk (IGC) Code proscribes design and construction standards of ships used in the transport of liquefied gasses and certain other substances. The review of the IGC Code was performed in conjunction with SOLAS) chapters II-2, VI and VII and submitted by Japan [ISWG/MASS 1/2/6].

Comments were provided indicating that the following sections (C) apply to MASS and do not prevent MASS operations but may need to be amended or clarified. These include in IGC chapter 16, "Use of cargo as fuel", which requires that fuel piping shall not pass through accommodation spaces, and in chapter 17, "Special requirements", for cargoes requiring a type 1G ship (i.e., some toxic cargoes) and "Chlorine" prescribe requirements for the provision of a space within the accommodation area to protect personnel against the effects of a major cargo release.

Additional comments were made pertaining the amendment to the requirements to manually operate the individual master valve of the gas fuel supply to each individual space containing a gas consumer(s) or through which fuel gas supply piping is run shall operate manually, that all rotating equipment utilized for conditioning the cargo for its use as gas fuel shall be arranged for manual remote stop from the engine room, and that a manually operated shut-off valve shall be fitted on the pipe of each gas-burner.

### 11.2.5 International Code on the Enhanced Programme of Inspections During Surveys of Bulk Carriers of Oil Tankers (ESP Code)

The International Code on the Enhanced Programme of Inspections During Surveys of Bulk Carriers of Oil Tankers (ESP) provides requirements for an enhanced programme of inspections of single and double-hull bulk carriers and oil tankers. The review of the ESP Code was submitted by Finland and supported by China [ISWG/MASS 1/2/12].

The results of the initial review determined that ESP Code concerns mainly surveys of ships and therefore requires no actions. However, the practical solution of having a survey report file with all supporting documents on board needs to be considered. Although comment was made that inspection procedures would need further consideration for all Degrees of Autonomy, it was pointed out that such inspections are commonly performed in collaboration with the technical department of the company.

### 11.2.6 International Code for the Safe Carriage of Packaged Irradiated Nuclear Fuel, Plutonium and High-Level Radioactive Wastes on board Ships (INF Code)

The International Code for the Safe Carriage of Packaged Irradiated Nuclear Fuel, Plutonium and High-Level Radioactive Wastes on board Ships (INF) supplements the International Atomic Energy Agency (IAEA) Regulations for the Safe Transport of Radioactive Material that are the principal regulations for radioactive transport. The review of this Code was submitted by Japan with support by China [ISWG/MASS 1/2/5].

The results of the initial review were agreed, in principle, with those of SOLAS, chapter 7. This includes the adoption of appropriate alternative safety measures to achieve the functionalities intended by the existing regulations and to establish emergency procedures to deal with conditions of leakage, spillage or fire involving cargoes.

### 11.2.7 International Code for Fire Safety Systems (FSS Code)

The International Code for Fire Safety Systems (FSS) provides international standards of specific engineering specification for fire safety systems required by SOLAS Chapter II-2. The review of the FSS Code was included in the review of Chapter II-2 submitted by Japan with support by China and the International Association of Classification Societies Ltd (IACS) [ISWG/MASS 1/2/3, Annex. 4 and ISWG/MASS 1/2/6].

Initial review comments include considering amending regulations or relegating to a separate instrument fire safety system control and operation including the presentation of indications and alarms to a location that is not onboard the ship. Additional comments were made regarding the unavailability of manual starting for diesel engines, the principles of means of escape while people are onboard for maintenance and other purposes; that fixed gas systems as well as fixed low-expansion fire-extinguishing systems currently were not automatic and require human intervention, the manipulation and use of a stop valve in each section of sprinklers, and the use of fixed pressure water-spraying and water-mist fire-extinguishing systems for IMO Degrees Three and Four Autonomy. Mention was also made with respect to fixed foam fire-extinguishing systems of a requirement for the capability for manual release with fixed high-expansion foam fire-extinguishing systems.

Amendments appear to be needed in FSS Code chapter 5, "Fixed gas fire-extinguishing systems", that prescribes when a device be provided which automatically regulates the discharge of the rated quantity of carbon dioxide into the protected area, that it shall be also possible to regulate the discharge manually; and in chapter 16, "Fixed hydrocarbon gas detection systems", that requires a mean of measurement of hydrocarbon gas with portable instruments.

## 11.2.8 Code of the International Standards and Recommended Practices for a Safety Investigation into a Casualty or Marine Incident (Casualty Investigation Code)

The Code of the International Standards and Recommended Practices for a Safety Investigation into a Casualty or Marine Incident (Casualty Investigation Code) provides for the investigation of marine casualties and incidents. The initial review was submitted by Finland with support from China [ISWG/MASS 1/2/12].

The initial review of results determined that since the Code concerns how to conduct marine safety investigations, most of its chapters (D) have no application to MASS operations. Some of the chapters concern ships, and that part of the Code (B) applies to MASS and does not prevent MASS operations and require no actions. Chapter 2, "Definitions", should include the term "location of control" as a substantially interested State which may be where the remote operator is located. This would be part of a broader discussion relating to responsibilities of the parties operating MASS where, in the case of an accident involving MASS the actions taken by the remote control center and personnel would need to be included in the investigation. It was also noted that the content of a marine safety investigation report should include the IMO Degree of Autonomy.

In chapter 12, "Obtaining evidence from seafarers", concerns were raised obtaining evidence from seafarers should be amended to include the remote control center and personnel and also to cover other relevant persons involved in operation. However, the initial review conclusion remains that when there are no seafarers involved, chapter 12 does not apply to fully autonomous ships since the goal of this chapter is to ensure the seafarers' human and legal rights.

### 11.2.9 International Maritime Solid Bulk Cargoes (IMSBC) Code

The International Maritime Solid Bulk Cargoes (IMSBC) Code facilitates the safe stowage and shipment of solid bulk cargoes and was reviewed in conjunction with SOLAS Chapter VI. The review of this chapter was submitted by Japan with support by China [ISWG/MASS 1/2/4Annex. 2, 1/2/6a].

Comments were provided indicating that sections of the Code "apply to MASS and do not prevent MASS operations but may need to be clarified". These clarifications include in section 3, "Safety of personnel and ship", whether additional locations for the instructions were necessary and how MASS would satisfy the requirements, In section 4, "Assessment of acceptability of consignments for safe shipment", how MASS would satisfy these requirements including: responsibilities of the shipper, who would be conducting the visual inspection, who would be conducting the sampling, and whether additional locations for the instructions were needed. In sections 5 and 7 of the Code, "Trimming procedures" and "Cargoes that may liquefy", how MASS would satisfy the responsibilities of the master who is required to take appropriate actions to prevent cargo shifting and potential capsize of the ship, consider seeking emergency entry into a place of refuge when a cargo shows an indication of liquefaction, and provisions for the ship's master's obligations during "carriage" in some individual schedules, e.g., for coal. Also of interest is section 8, "Test procedures for cargoes that may liquefy", as to whom would conduct the can test of paragraph 8.4. In section 10, "Carriage of solid wastes in bulk", there is a need to reassign duties assigned to the Master in paragraph 10.9; and section 11, "Security provisions", should be reassessed in terms of vessels with no seafarers on board. Additional comments were provided on Appendix 1 relative to IMO Degrees of Autonomy Three and Four as to how MASS would satisfy the continuous cargo monitoring for a variety of cargoes without seafarers on board, and where certificates must be located.

### 11.2.10 International Code for the Safe Carriage of Grain in Bulk (International Grain Code)

The International Code for the Safe Carriage of Grain in Bulk (International Grain Code) facilitates the safe stowage and shipment of bulk grain, and was reviewed in conjunction with SOLAS Chapter VI. The review of this chapter was submitted by Japan with support from China [ISWG/MASS 1/2/4Annex. 3].

Comments were provided indicating (C) applicability to MASS and do not prevent MASS operations including Part A regulation 3, "Document of authorization", where the location of documentation should be reassessed, and regulation 18, "Securing with wire mesh", where clarification was needed on how the requirement to inspect the lashing during the voyage would be fulfilled. Additional comments were made regarding regulation 4, "Equivalents", citing a need to change define the LSA regulations for MASS; and regulation 17, "Strapping or lashing", citing the necessity to consider how the requirement to inspect the strapping during the voyage would be fulfilled with no seafarers on board.

### 11.2.11 Code of Safe Practice for Cargo Stowage and Securing (CSS) Code

The Code of Safe Practice for Cargo Stowage and Securing (CSS) considers measures to deal with hazards from a combination of longitudinal, vertical and transverse motions and the forces created by these accelerations by properly securing proper stowage and securing cargoes on board and to reduce the amplitude and frequency of ship motions. This code was reviewed in conjunction with SOLAS Chapter VI and the review of this chapter was submitted by Japan with support from China [ISWG/MASS 1/2/4Annex. 3].

Comments were provided indicating the Code (C) applies to MASS and do not prevent MASS operations yet require assessment as specifically pertaining to paragraph 1.9.2 in the provisioning of cargo information to the master.

### 11.2.12 International Convention on Standards of Training, Certification and Watchkeeping for Fishing Vessel Personnel (STCW-F Convention)

The International Convention on Standards of Training, Certification and Watchkeeping for Fishing Vessel Personnel (STCW-F) 1995 addresses watchkeeping and minimum training and certification requirements for crews of seagoing fishing vessels of 24 meters or more in length. The review of this convention, submitted by Japan with support from New Zealand and Spain, considered the result obtained for the STCW 1978 convention (described in Section 11.1.4) to maintain consistency [ISWG/MASS 1/2/10].

Most of the articles and regulations were identified as (D) having no application to MASS as they concern the responsibilities of Administrations. The exceptions are Article 3 (Application), Article 6 (Certification) and Article 8 (Control) for Degree of Autonomy Three; and all regulations in chapters II, III and IV that were identified (C) as applying to MASS, do not prevent MASS operations but may need amendment or clarification for MASS of IMO Degree of Autonomy Three, considering the case that remote control operators were within the scope of the Convention, and for IMO Degree of Autonomy One and Two considering the necessity to introduce new technologies or automated processes.

Other comments cited a need to add some new terms related to MASS and to extend the scope of the Convention over the remote control operators of MASS. The establishment of a legal framework for the remote operator would be a significant gap between the Convention and MASS operation and the regulations may need to be reconsidered if remote control operators were recognized as seafarers covered by the Convention. Moreover, all regulations of chapters II, III and IV would contain gaps for MASS operation under Degrees of Autonomy One and Two considering the necessity to add relevant knowledge, understanding and proficiency when new technologies or automated processes are introduced on board.

### 11.2.13 International Convention on Load Lines (LL Convention)

The International Convention on Load Lines (LL) 1966 makes provisions for determining the freeboard of ships by subdivision and damage stability calculations. The initial review was submitted by India with support from China and Liberia [ISWG/MASS 1/2/18].

The results of the initial review determined that the concept of assigning freeboards and Load Line Marks remains relevant for safety of all degrees of MASS. Therefore most regulations (C) remain applicable to all categories of MASS, with amendments being required for categories of MASS without crew on board (degrees three and four) in relation to activities requiring manual intervention/presence of crew on board. For Autonomy Degrees One and Two it is assumed that these ships will have qualified seafarers on board who can perform load line-related functions for maintaining the weather and watertight integrity of the vessel (i.e., closing of vent covers, piping system shut-off valves, etc.). Such provisions are therefore categorized as applying to MASS and do not prevent MASS operations and require no actions for MASS operating under IMO Degrees One and Two Autonomy.

Concerning the Articles of 1988 Protocol of the Load Line Convention these are considered to be applicable to all degrees of MASS with the understanding that they will be considered as New Ships, under the Convention.

### 11.3 LITTLE OR NO SIGNIFICANCE TO REGULATIONS TO SUPPORT MASS OPERATIONS

The following instruments are expected to require few or no amendments related to the implementation of MASS.

### 11.3.1 IMO Instruments Implementation (III Code)

The IMO Instruments Implementation (III) Code addresses the effective and consistent global implementation and enforcement of IMO instruments. The initial review was submitted by India with support from China and Liberia [ISWG/MASS 1/2/18].

The results of the initial review determined that provisions of the Code are relevant to all degrees of MASS. Some parts of the Code, such as obligations of Flag, Coastal and Port States, may need revision to account for additional/alternate/equivalent responsibilities in relation to MASS operating in Degrees Two, Three and Four Autonomy.

### 11.3.2 International Code on Intact Stability (IS Code)

The International Code on Intact Stability (IS) stipulates regulations for intact stability for all types of ships covered by IMO instruments. The initial review was submitted by India with support from China and Liberia [ISWG/MASS 1/2/18].

The results of the initial review determined Part A of the IS Code is considered relevant to all degrees of MASS. For MASS operating at degrees two and three automation references to "master" used in sections of Part A are understood to apply to whoever is in charge of the vessel at any point in time. The remote operator or the onboard seafarer responsible for taking over responsibilities can be considered as an authority equivalent to "master". This needs to be clarified in the relevant provisions. For MASS of Degree Four Autonomy, it is necessary to identify and assign responsibility to an equivalent authority, before this category of MASS becomes operational.

### 11.3.3 International Code for Application of Fire Test Procedures (FTP Code)

The International Code for Application of Fire Test Procedures (FTP) contains fire test procedures for fire safe construction and materials used on ships. The initial review was submitted by Japan [ISWG/MASS 1/2/18].

The results of the initial review were unanimously agreed that the Code (B) applies to MASS, do not prevent MASS operations and require no actions.

### 11.3.4 International Life-Saving Appliance (LSA Code)

The International Life-Saving Appliance (LSA) Code covers specific technical requirements for life-saving appliances. The initial review was submitted by Belgium and the Netherlands with support from China [ISWG/MASS 1/2/11].

The review of the LSA Code revealed that the requirements therein relate to the performance and construction of LSA, not to the carriage requirement, and can be complied with regardless the degree of MASS operation.

### 11.3.5 Code for Recognized Organizations (RO Code)

The Code for Recognized Organizations (RO) clarifies the responsibilities of organizations recognized as ROs for a flag State. The initial review was submitted by Finland with support from China [ISWG/MASS 1/2/12].

The initial review determined that the RO Code concerns Recognized Organizations and therefore has no application to MASS.

### 11.3.6 International Convention on Tonnage Measurement of Ships (Tonnage Convention)

The International Convention on Tonnage Measurement of Ships (Tonnage Convention) 1969 introduced a universal tonnage measurement system for merchant ships providing for gross and net tonnages. The initial review was submitted by Liberia [ISWG/MASS 1/2/17].

The initial review determined that for IMO Degrees of Autonomy One, Two, Three and Four, all articles and regulations are assessed (B) to apply to MASS and do not prevent MASS operations and require no actions with the exception of Article 2 and Regulation 2. Both Article 2 ("Definitions") and Regulation 2 ("Definitions of Terms used in the Annexes") relate to definitions, and in particular the definition of the Master, crew and passenger needs to be clarified in the context of MASS operation.

Also, comment was made regarding Regulation 6 (Calculation of Volumes) for IMO Degrees of Autonomy Two, Three and Four, where there may need to be consideration for whether MASS require new or additional areas to be measured to calculate volume. This comment is considered relevant; however, it is suggested the issue raised should be addressed in further discussion. For the purpose of this first step it is considered that Regulation 6 could in principle still "(B) apply to MASS and do not prevent MASS operations".

### 11.3.7 International Convention for Safe Containers (CSC Convention)

The International Convention for Safe Containers (CSC Convention) 1972 is intended to make it possible to maintain a high level of safety of human life in the transport and handling of shipping containers by providing uniform safety standards. The review of the CSC Convention was submitted by Japan and supported from Finland [ISWG/MASS 1/2/9].

The result of the initial review indicated that the requirements of all articles of the Convention (B) apply to MASS, do not prevent MASS operations and require no actions regardless of Degree of Autonomy. Comments received during the commenting period indicated a belief that there exists no relationship between CSC and the various proposed modes of operating MASS, but there may exist portions of the Cargo Stowage and Securing Code (CSS Code) that are more applicable to this regulatory scoping exercise.

## 11.4 INSTRUMENTS NOT YET CONSIDERED

The following instruments have not yet been considered relative to the implementation of MASS.

### 11.4.1 International Convention for the Prevention of Pollution from Ships (MARPOL Convention)

The International Convention for the Prevention of Pollution from Ships (MARPOL Convention) 1973 as modified by the Protocol of 1978 relating thereto and by the Protocol of 1997 includes regulations aimed at preventing and minimizing pollution from ships from both accidents and from routine operations.

### 11.4.2 International Code of Safety for High-Speed Craft (HSC Code)

The International Code of Safety for High-Speed Craft (HSC), 1994 and 2000, incorporates a comprehensive set of requirements for high-speed craft that include, among others, air-cushion vehicles such as hovercraft and hydrofoil boats.

### 11.4.3 International Code of Safety for Ships Using Gasses or Other Low-Flashpoint Fuels (IGF Code)

The International Code of Safety for Ships Using Gasses or other Low-Flashpoint Fuels (IGF) provides regulations to minimize the risk to ships, their crews and the environment given the nature of the fuels involved.

### 11.4.4 Code for the Construction and Equipment of Ships Carrying Dangerous Chemicals in Bulk (BCH Code)

The Code for the Construction and Equipment of Ships Carrying Dangerous Chemicals in Bulk (BCH) provides international standards for the safe carriage of dangerous and noxious chemicals in bulk by prescribing the construction features of ships.

### 11.4.5 Special Trade Passenger Ships Agreement (STP Agreement)

The Special Trade Passenger Ships Agreement (STP Agreement) 1971 considers safety requirements for the carriage of large numbers of unberthed passengers in special trades.

### 11.4.6 Protocol on Space Requirements for Special Trade Passenger Ships (Space STP Protocol)

The Protocol on Space Requirements for Special Trade Passenger Ships (Space STP) 1973 develops technical rules covering the safety aspects of carrying passengers under the Special Trade Passenger Ships Agreement.

### 11.4.7 Code of Safety for Nuclear Merchant Ships

The Code of Safety for Nuclear Merchant Ships is a guide to Administrations on Internationally accepted safety standards for the design, construction, operation, maintenance, inspection, salvage and disposal of nuclear merchant ships.

### REFERENCES

IMO CONV. Key IMO Conventions. http://www.imo.org/en/About/Conventions/ListOfConventions/Pages/Default.aspx.

International Convention for the Safety of Life at Sea (SOLAS) 1974. International Maritime Organization. http://www.imo.org/en/About/Conventions/ListOfConventions/Pages/International-Convention-for-the-Safety-of-Life-at-Sea-(SOLAS),-1974.aspx.

ISWG/MASS 1/2. Summary of Results of the First Step of the RSE for SOLAS Chapter IX and the ISM Code in Relation to the Safe Operation of Maritime Autonomous Surface Ships. Norway. Intercessional Working Group on Maritime Autonomous Surface Ships. Agenda Item 2. 24 July 2019.

ISWG/MASS 1/2/2. Summary of Results of the First Step of the RSE for the International Convention on Maritime Search and Rescue (SAR). France and Spain. Intercessional Working Group on Maritime Autonomous Surface Ships. Agenda Item 2. 28 July 2019.

ISWG/MASS 1/2/3. Summary of Results of the First Step of the RSE for SOLAS Chapter II-2 and Associated Codes. Japan. Intercessional Working Group on Maritime Autonomous Surface Ships. Agenda Item 2. 29 July 2019.

ISWG/MASS 1/2/4. Summary of Results of the First Step of the RSE for SOLAS Chapter VI and Associated Codes. Japan. Intercessional Working Group on Maritime Autonomous Surface Ships. Agenda Item 2. 29 July 2019.

ISWG/MASS 1/2/5. Summary of Results of the First Step of the RSE for SOLAS Chapter VII and Associated Codes. Japan. Intercessional Working Group on Maritime Autonomous Surface Ships. Agenda Item 2. 29 July 2019, pp. 2, 3.

ISWG/MASS 1/2/6. Findings and Common Issues Identified in the Initial Review of Chapters II-2, VI and VII of the Annex to SOLAS 1974 and the Associated Codes. Japan. Intercessional Working Group on Maritime Autonomous Surface Ships. Agenda Item 2. 29 July 2019.

ISWG/MASS 1/2/7. Summary of Results of the First Step of the RSE for SOLAS Chapter XII and Associated Standards. Japan. Intercessional Working Group on Maritime Autonomous Surface Ships. Agenda Item 2. 29 July 2019.

ISWG/MASS 1/2/8. Summary of Results of the First Step of the RSE for SOLAS Chapter XIII and Associated Standards. Japan. Intercessional Working Group on Maritime Autonomous Surface Ships. Agenda Item 2. 29 July 2019.

ISWG/MASS 1/2/9. Summary of Results of the First Step of the RSE for CSC 1972. Japan. Intercessional Working Group on Maritime Autonomous Surface Ships. Agenda Item 2. 29 July 2019.

ISWG/MASS 1/2/10. Summary of Results of the First Step of the RSE for STCW-F 1995. Japan. Intercessional Working Group on Maritime Autonomous Surface Ships. Agenda Item 2. 29 July 2019.

IWSG/MASS 1/2/11. Summary of Results of the First Step of the RSE for SOLAS Chapter III and the LSA Code. Belgium and the Netherlands. Intercessional Working Group on Maritime Autonomous Surface Ships. Agenda Item 2. 30 July 2019.

ISWG/MASS 1/2/12. Summary of Results of the First Step of the RSE for SOLAS Chapter XI and Related Codes. Finland. Intercessional Working Group on Maritime Autonomous Surface Ships. Agenda Item 2. 31 July 2019.

ISWG/MASS 1/2/13. Summary of Results of the First Step of the RSE for SOLAS Chapter XI-2 and the ISPS Code. Finland. Intercessional Working Group on Maritime Autonomous Surface Ships. Agenda Item 2. 31 July 2019.

ISWG/MASS 1/2/14. Summary of Results of the First Step of the RSE for SOLAS Chapter XIV and the Related Polar Code. Finland. Intercessional Working Group on Maritime Autonomous Surface Ships. Agenda Item 2. 31 July 2019.

ISWG/MASS 1/2/15. Summary of Results of the First Step of the RSE for SOLAS Chapter IV. Turkey. Intercessional Working Group on Maritime Autonomous Surface Ships. Agenda Item 2. 1 August 2019.

ISWG/MASS 1/2/16. Summary of Results of the First Step of the RSE for SOLAS Chapter V. China and Singapore. Intercessional Working Group on Maritime Autonomous Surface Ships. Agenda Item 2. 2 August 2019.

ISWG/MASS 1/2/17. Summary of Results of the First Step of the RSE for the International Convention on Tonnage Measurement of Ships, Liberia. Intercessional Working Group on Maritime Autonomous Surface Ships. Agenda Item 2. 2 August 2019.

ISWG/MASS 1/2/18. Summary of Results of the First Step of the RSE for LL 66, PROP 68, IS Code Part A and III Code. India. Intercessional Working Group on Maritime Autonomous Surface Ships. Agenda Item 2. 2 August 2019.

ISWG/MASS 1/2/19. Summary of Results of the First Step of the RSE for the International Regulations for Preventing Collisions at Sea (COLREG). Marshall Islands. Intercessional Working Group on Maritime Autonomous Surface Ships. Agenda Item 2. 2 August 2019.

ISWG/MASS 1/2/20. Summary of Results of the First Step of the RSE for STCW Convention and Code. United States. Intercessional Working Group on Maritime Autonomous Surface Ships. Agenda Item 2. 2 August 2019.

LEG 106/8/1. Outcomes of MSC 99 and MSC 100 regarding MASS. Note by the Secretariat. Legal Committee. 106th Session. Agenda Item 8. 11 January 2019, Annex 2.

MSC 101/5. Status Report – Progress of the Regulatory Scoping Exercise. Note by the Secretariat. Maritime Safety Committee 101st Session, Agenda Item 5. 1 April 2019, Annex 1–3.

MSC/Circ 1432. Revised Guidelines for the Maintenance and Inspection of Fire Protection Systems and Appliances. International Maritime Organization. 31 May 2012.

# Chapter 12

# Future Directions of MASS

The concept of Maritime Autonomous Surface Ships (MASS) is presently in its infancy, with independent research and development efforts currently underway and test beds and prototype systems being developed but with little crossover in terms of knowledge and results sharing. The International Maritime Organization (IMO) has concluded its regulatory scoping exercise and the task of adapting existing regulatory instruments or creating new instruments to fulfill MASS requirements has just begun, with its completion still years away. The approval of Interim Guidelines for MASS Trials by the IMO Maritime Safety Committee (MSC) is a good first step, and many issues are currently being addressed in known problem areas having to do with MASS in closed environments and their interaction with conventional vessels while underway [MSC 101/5/5]. However, there are many issues that are presently not being addressed as it may be premature to do so. This includes identifying specific details of how regulations should be changed before determining what regulations are even applicable to MASS. Other issues are related to new technologies for which no present maritime regulations exist and where guidance is needed from other industries more experienced with autonomous systems. More so are the unintended consequences that may result from the inclusion of cutting-edge new technology in MASS that has never before been used and new frontiers are being explored.

This chapter can be considered a summary of some of the many significant topics that need to be considered as MASS projects extend into the future. It attempts to identify some of the issues that must be considered before MASS may be deemed sufficiently mature to enter the mainstream of the maritime industry. Some topics already addressed to some extent in this book are also included with editorial comments and opinion not expressed in the technical discussion to provide further elaboration and to emphasize their significance. The list is far from comprehensive and it is expected to rapidly change over the next few years as experience is gained, lessons are learned, problems are solved and new issues arise.

## 12.1 DEMONSTRATED COMPETENCY OF MASS

The maritime professions spanning many disciplines and ship's departments require seafarers of all ranks and nationalities to be trained and certified for service on modern seagoing ships in accordance with the latest changes and amendments to the International Convention on Standards of Training, Certification and Watchkeeping (STCW). The assessment procedures require verification that members of the crew who are required to be competent do in fact possess the necessary skills [STCW/CONF 2/34]. The express goal of MASS is to facilitate the remote and/or autonomous operation of vessels with varying amounts of human participation, including fully autonomous operation at IMO Degree Four where human participation is expected to occur only if the system fails or prompts for human intervention. Therefore, since MASS is expected to perform without humans onboard, the question is posed as to what standards of competency should be put into place to ensure that MASS reasoning system software, firmware and hardware components (the brains of MASS) are adequately trained and certified as being competent for the functions being performed. Another question is whether performance evaluated as being at least equal to human performance is sufficient, realizing that humans constantly make mistakes based upon factors such as lack of training, inexperience in the practical application of training, reasoning with uncertainty as well as distractions and other unforeseen circumstances. Or should machines be held to other standards that humans can never be expected to achieve: a defacto "double standard" that would ensure the greatest safety and elimination of risk. However, such a standard may not ultimately be possible.

The primary consideration for this issue is the fact that artificial intelligence technologies, especially deep learning neural networks, are presently inherently unverifiable and therefore unable to be proven or even objectively demonstrated as being accurate and reliable using other than circumstantial evidence. AI-based systems lie at the heart of MASS in identifying and recognizing objects (e.g., aids to navigation and other vessels), trends in data that may indicate potential conflicts in course that may lead to collision or allision, or even discrepancies in data that may adversely affect decision making or outcomes. This is due mainly to a lack of visibility into the internal working mechanisms and architectures of neural networks. These circumstances are expected to change in coming years as artificial intelligence implementations become widespread and more transparent architectures are developed. In the meantime, methods must be created to ensure that the reasoning systems used on MASS can be validated using empirical methods that take advantage of the huge volumes of data expected to be generated and that the results obtained are predictable and traceable.

## 12.2 FITNESS FOR DUTY

A trained seafarer that is fully qualified for their position with all of the required technical and medical certifications and endorsements is still not fit if they are unable to perform their assigned tasks. The fitness for duty evaluation considers physical capabilities for hearing and vision as well as physical and mental states and impairments that can affect performance of routine and emergency duties [ILO 2011]. Likewise, reasoning systems for use on MASS and in remote control centers (RCCs) determined as being competent as described in the previous paragraph must still be certified as being able to manage the specific risks and impacts associated with fielding the system in actual use. Competency is considered the successful result of a correct and verifiable development process to achieve stated goals, while fitness for duty considers the ongoing ability of the reasoning system to continue to function properly.

Analogies to seafarers can be made to hearing and vision with corresponding MASS sensor capabilities in determining which and how many sensor inputs and data may be critical and their loss or compromise yet still be considered acceptable. Likewise, a seafarer having all required immunizations needed for sea service can correspond to MASS reasoning systems being updated with the latest security software patches to protect them against hacking and the introduction of known viruses.

Fitness for duty considerations have stymied software developers for decades yet the quality control and assurance methods for AI-based reasoning and especially MASS applications have yet to be developed. The implementation of digital twin simulation to detect and analyze in real time potential differences in reasoning outcomes using the same sensor stimulus as discussed in Chapter 5 can greatly assist in achieving and maintaining this goal. Although these criteria have been addressed in flight-critical aircraft and spacecraft components, the required practices are not generally in accordance with the culture of the maritime industry. This is an area where maritime enterprises can benefit from the experiences of other industries more advanced in autonomy.

## 12.3 SECURITY

The great Achilles' heel for MASS will be in the area of cyber security in attempting to prevent and overcome attempts at sabotage and hijacking vessels while at sea and also to deter hacking by preventing the implantation of viruses and taking advantage of "back doors" and other vulnerabilities inherent to networked computer-based systems. Such attempts will become more frequent and widespread in the future as the stakes get higher and the potential increases for nefarious undertakings that can result in mass casualties and economic disaster, and MASS will be a prime target.

Despite all attempts at minimizing such problems, the innovative preventive measures and solutions produced to solve today's problems will barely keep pace with adversaries who are just as creative and motivated to formulate new means and methods for attack. A good analogy to this problem lies in the efforts presently being undertaken to secure elections from hacking in the United States. Despite millions of dollars in expenditures to develop secure voting systems with the latest hardware and software solutions and taking precautions such as not connecting voting machines directly to the Internet, the vast majority of the 10,000 election jurisdictions nationwide that will be acquiring these new systems use older software to support them to create the ballots, program the voting machines, tally votes and report vote counts [Abdollah 2019]. Most of these voting systems are based upon Windows 10, which has many of the latest software fixes and security fixes and can be updated frequently. However, many of the supporting systems use Windows 7 (and even earlier versions), which reached its "end of life" on 14 January 2020, meaning that Microsoft will no longer routinely provide technical support or produce software patches to fix software vulnerabilities. While Microsoft may provide future support for a fee to provide continued Windows 7 security updates through 2023, such an approach will be piecemeal in coverage and inferior in results. This has resulted in many election districts returning to the use of paper ballots that are verifiable and traceable.

Extend this scenario to MASS where the vessels and remote control centers (RCC) utilize potentially hundreds (or thousands) of different software automation and support systems created by hundreds of developers and installed by even more vendors, each of which are responsible for their own security measures and software maintenance guidelines. Add to this the risks associated with broadband links over which the huge volumes of sensor and other data along with command and control instructions are continuously exchanged that are susceptible to interception and hacking. Further, satellite-based Global Navigation Satellite System (GNSS), Automatic Identification System (AIS), Radar and other space and terrestrial technologies supporting the critical functions of navigation are subject to spoofing, hacking and denial of service attacks that can directly endanger and isolate MASS from the RCC while underway. These vulnerabilities can increase MASS susceptibility to hijacking and, if under the control of the wrong entities, to make these vessels into tools for destruction. One example of how MASS can be used in this way is that of a 10 meter (32 foot) remotely controlled vessel laden with explosives found in the path of the British destroyer HMS *Duncan* [Hughes 2019]. These combined risk elements will surely challenge safe and secure implementations of future MASS operations.

## 12.4 ENVIRONMENTAL CONCERNS

Digitalization of the shipping and other industries is enhancing both ship and port operations by creating new tools that can provide greater insight into ship and port operations and can bring enhanced efficiencies into their operations. While few dispute the benefits that may be achieved through such endeavors, there are several drawbacks to these approaches and the technologies involved that do not make the headlines yet should raise alarms and be seriously considered as MASS operations move closer to fruition.

### 12.4.1 5G Broadband Technology

The first alarm has to do with the many thousands of 5G satellite and terrestrial communication nodes that will make possible ubiquitous broadband communications beginning in the early 2020s. Much concern has been raised about the effects of cellular and other radio frequency (RF) technologies on human health, and there remains a great deal of dispute and misinformation on this subject leading to an inability to achieve clear consensus on either the problem or potential solutions. However, it is my contention that this is merely the tip of the iceberg when it comes to the potential negative impact of these technologies on the environment with the Earth representing the *Titanic* and the world's population its passengers and crew. Planned 5G communications use entirely new technologies that operate at microwave frequencies. The problem lies in the amount of locally emitted RF energy by terrestrial 5G nodes that continuously bathe the environment with microwave radiation. The combined overall strength and intensity of these emissions must necessarily be greater than present-day cellular technologies since range is much reduced at their much higher operating frequencies (~5 GHz for 5G vs. ~2 GHz for cellular) and affected by weather (e.g., heavy rain). This will result in a need to install many terrestrial 5G nodes to achieve the same geographical coverage as one present-day cellular tower [RHR 2019].

5G satellite systems pose a similar danger for an entirely different reason. These systems provide communications using narrow beams of microwave energy radiated from space aimed at the Earth's surface. It is estimated that by the mid-2020s there will be tens of thousands of 5G communications satellites configured into several constellations in low Earth orbit placed there by SpaceX, Boeing, OneWeb, Spire Global and other companies in this rapidly growing industry [World Health 2019]. This will result in worldwide 5G coverage without gaps, even in the polar regions where broadband communications have slowly been introduced over the past few years at lower frequencies by Inmarsat and other satellite communications providers.

The combined effects of manmade terrestrial and satellite 5G microwave radiation warming the Earth and its population have received little study and their effects are presently unknown. It seems to me this should gain greater significance on someone's agenda before proceeding further with widespread implementation. However, the 5G revolution is presently well underway, with billions of dollars already invested.

### 12.4.2 Contribution of Greenhouse Gas to the Environment

A French study published in 2019 concluded that online videos are said to generate 60% of world data traffic, with the energy used to accomplish this estimated to contribute more than 300 million tons of carbon dioxide emissions per year [Efoui-Hess 2019]. The intensive volume of imagery and other data associated with MASS, remote control centers and smart ports will swamp present-day maritime data rates and add to this problem. Put into perspective, the aggregated environmental footprint of digital technologies on a global scale is purported to represent close to 4% of worldwide carbon emissions, which is claimed to be more than civil air transport, with their impact increasing by 8% per year. Without commenting on the methodology used by the researchers in the study or the conclusions they have reached, and especially their proposed regulation-intensive approach to solving the problem, it is noteworthy that little consideration has been given to this topic in the widespread discussion on MASS in the news and literature.

New propulsion technologies and fuels are being actively sought for use by MASS, with many claims of potential reduced emissions being made. The use of electricity and batteries presently appears to be a favored solution, especially for short-distance voyages such as those considered for *Yara Birkeland* and several ferries discussed in this book. One must keep in mind that electrical power provides clean, emission-free propulsion at the point of use. However, the power needed to charge batteries and to operate electromagnetic mooring, docking and other systems is not emission-free at the point of its generation. Likewise, the power generated to support digital communications comes with a cost in terms of economics and emissions.

There should be a discussion among the maritime and environmental communities considering both sides of the ledger when it comes to claims of MASS for environmental impact. This discussion should include the cost and overall carbon footprint for energy production to power MASS and drive its communications lifeline. Also should be considered are the effects associated with the potential for use of heavy metals and dangerous materials in the batteries and other technologies being proposed for use in MASS systems and architectures along with the costs for disposal and recycling throughout their life cycle.

## 12.5 SMART PORTS

At a recent maritime conference the topic of MASS and smart ports came up in the discussion to the effect that a "smart ship can do very little in a dumb port" [*TransNav* 2019]. The digitalization of ports and terminals through the adoption of modern technologies is leading to more efficient operations throughout the logistics chain and promoting better safety. The technologies used to implement digitalization include interconnected computer platforms combined with cloud-based services, a wide range of sensors and Internet of Things (IoT) devices, augmented and virtual reality, blockchain technology and big data enterprises facilitates data sharing among local authorities, intermodal shipping, railroad and trucking lines as well as storage providers.

Digital services can provide seamless integration of MASS and port operations by exchanging information regarding cargoes and transportation assets to afford better scheduling through preplanning movements at the quayside and in the terminal. Benefits are also gained by increased sensor use to detect and identify infrastructure deficiencies and bottlenecks, optimizing cargo operations and improving payments; to provide early warning of potential delays and interruptions; and to reduce energy consumption. MASS outfitted to a high degree with sensors and instruments can form an integral part of the port facility architecture to achieve these goals as an Internet of Things (IoT) node, providing enhanced compositional information regarding the vessel's cargo along the entire chain of custody. Much of these data can be packaged and marketed as new, value-added products useful for optimizing the operations of many of the companies and organizations involved in both local and long-distance transportation, to enhance port and MASS security and to improve terminal safety. Research into the digital integration of MASS with port and terminal systems is warranted and should be considered as part of future projects.

## 12.6 AIDS TO NAVIGATION

The subject of aids to navigation (ATON) was touched on briefly in Chapter 6 with respect to how MASS will utilize existing ATON infrastructure to guide their transits. There also exists an opportunity to devise new ATON designs that can take advantage of machine vision devices that future conventional vessels will use to assist the bridge watch during navigation and MASS will rely upon to provide geographical fixes to supplement GNSS and other positioning resources. The capability of machine vision to observe, detect and identify objects with accuracy that is at least comparable to, if not better than, human ability, to watch constantly without becoming tired, bored or distracted; or require relief after a few hours makes them a vital addition to the watchstander tool kit. The need for such automated capabilities is without question, as evidenced by countless investigations stating that watchstander disorientation was a significant contributing factor for an accident. The problem continues in modern times for crews of even the latest and best-equipped civilian and naval vessels.

There exists opportunities for research in two directions. The first involves the integration of new characteristics into existing ATON designs for buoys and day beacons, lights and ranges and even active ATON such as Radar beacons and AIS-ATON. New characteristics include machine-readable signage containing additional encoded information beyond the buoy number, yet still appearing in all aspects like a conventional buoy. Additional examples include multispectral paint visible under various wavelengths of light beyond the visible spectrum, and even passive Sonar targets that may be affixed to chains and sinkers to enhance Sonar detection of ATON that can be readily integrated into existing ATON systems, both fixed and afloat, with no or minimal modification and without interfering with present form, fit or function. Other considerations include the expansion of active, smart transponder use to provide identification under interrogation by Radar, Sonar, radio, Wi-Fi and other sources. Such transponders could be similar to those widely in use to aid in road toll collections on major highways and bridges worldwide.

The second area for research involves the development of virtual aids to navigation (VATON) that do not require physical infrastructure such as VHF radio as needed for present-day AIS-ATON [Wright 2019; Wright and Baldauf 2016b]. These virtual devices are georeferenced to the environment to assure positional accuracy and exist as two-dimensional waypoints (latitude and longitude) for surface ships, and even three-dimensional waypoints (latitude, longitude and depth) to guide submarines and autonomous undersea vehicles (AUVs). VATON may be placed anywhere on the planet and are especially useful in polar and remote regions where physical ATON are too costly to place or maintain. A simple single-beam echosounder mandated for use on all but the smallest of vessels provides the means to verify VATON positioning and to detect changes in bottom topography since the last hydrographic survey that may indicate a change in position may be needed to watch best waters. This same technology can be used to routinely verify the position of physical ATON by any vessel passing by that can be reported to ATON authorities, enhancing the safety of Federal and private ATON systems.

## 12.7 MASS OPERATOR COMPLACENCY

An incident involving an Uber autonomous vehicle striking and killing a pedestrian crossing the street at a properly marked crosswalk with a traffic light should raise warning flags to all those concerned with MASS. This incident occurred in March 2018 in Tempe, Arizona, while the required safety and emergency backup driver was seated in the vehicle [Wakabayashi 2019]. Rather than criticizing the automation technology or disparaging the safety driver for allowing such an accident to happen, my reaction to the news was along the lines of "of course such incidents are going to happen" and "why didn't they anticipate this?"

Pairing an experimental technology in an uncontrolled setting with a human who is expected to sit idly watching the vehicle for hours at a time, day after day, week after week, without having to intervene or do anything but watch is an impossible task. This death was caused by a combination of technical and human failures. Should the reasoning system have seen, properly identified and reacted to the pedestrian? Absolutely! Should the safety driver have detected the failure and reacted to prevent the incident? Yes, and maybe! Should the vehicle have been allowed to transit the public roadways while using unverified and experimental technology? Maybe not!

On its face it appears that the onboard navigation reasoning system encountered a situation for which it was unprepared. The exact cause will undoubtedly be determined as a result of the accident investigation. Whether this cause represents the only possible scenario for which the reasoning system was not prepared seems extremely unlikely. Ideally, the safety driver would have taken over and prevented the incident. However, the safety driver is human, with all its benefits and limitations and is subject to inattention, boredom, distraction, sleepiness and myriad other potential shortfalls. Knowingly or not, the driver was allowed to enter the situation without properly appreciating the consequences of failure. This leaves the remaining part of the human equation to the regulatory authorities.

The question not being addressed that is equally applicable to both road vehicles and MASS is how to reconcile the outcomes of having to make the best choice of several bad and worse choices. Should a vehicle swerve into oncoming traffic and risk the likelihood of an accident with multiple injuries and deaths rather than stay in lane with a high certainty of killing a pedestrian? Likewise, should an automated ship ground itself and risk loss of property and great ecological damage rather than continue on its course with a high degree of certainty of collision with another vessel?

The IMO must be given credit for recognizing the need for and completing the regulatory scoping exercise as a first step toward regulation and oversight. The next step of implementing a regulatory mechanism must be well thought out and promptly achieved. Caution must be taken on the part of testbeds and test areas to ensure adequate safety is assured prior to unleashing MASS on the public waterways and high seas.

## 12.8 IS IMO DEGREE 4 FULL AUTOMATION FOR MASS ETHICAL, OR EVEN POSSIBLE?

This question is at the heart of the problem for which much research, investigation, reporting and sharing of information, and attempts at oversight are focused. There are vested interests that would like to proceed relatively unhindered with experimentation and testing, while others would prefer the entire concept of MASS go away and no more be mentioned.

The Uber incident of the previous paragraph lends credence to the notion that the implementation of new concepts and new solutions to old problems is not without risk. A reduction of risk to help prevent such incidents can be achieved through the use of digital twin simulation to see exactly what MASS will do in real time using actual live sensor data when encountering traffic, obstacles and unexpected events while under the command of onboard seafarers. Such measures were not taken with Uber and are not the norm in vehicle automation trials. Had this been previously accomplished using a large fleet of vehicles under direct human control with digital twin simulation being accomplished with live data, such scenarios may possibly have been detected and the reasoning software updated accordingly.

This begs the question of whether IMO Degree Three remote operation should be allowed without the potential for IMO Degree Four Full Automation operation by the MASS itself to automatically take over navigation in the event the remote MASS operator appears to make a mistake. Indeed, what may appear to the MASS reasoning system to be an operator mistake may actually be a third party attempt to highjack the vessel.

The verification and validation of MASS technologies must not be rushed, adequate facilities must be established onboard MASS and at the remote control centers, and proper training must be established and implemented. This process will, and should, take years. However, the direct benefits to MASS and the indirect benefits to conventional ships and the shipping industry as a whole through derivative products and services will be more than worthwhile.

## 12.9 SITUATIONAL AWARENESS BELOW THE WATERLINE

In previous chapters there is discussion regarding improving overall MASS situational awareness, and especially for the environment below the waterline using an assortment of subsea sensors (see Section 6.3.2.2 in Chapter 6). Recent developments in side-scan and forward-looking navigation Sonar technology have transformed our ability to detect and identify underwater features and to avoid hazards to navigation [Wright and Baldauf 2014]. At present there are no mandated requirements for vessels to carry such equipment. However, significant benefits enhancing safety of ship navigation may be accrued from their use. This is especially important in waters that are poorly surveyed or subject to shoaling and shifting sea beds and where objects and hazards may exist that are not depicted on navigation charts which can lead to groundings or allision.

Side-scan Sonar creates high-resolution imagery of the sea floor and is commonly used for hydrographic survey and marine archaeology to detect objects and debris. Of particular interest to MASS operations is its ability to identify significant objects and bottom features not present on navigation charts that can be used as landmarks to supplement land and sea-based aids to navigation in determining position. This can be accomplished without having to rely on GNSS or radio-based (e.g., eLoran) electronic positioning system resources that are subject to interference, spoofing and denial of service attacks. Side-scan imaging capability is presently being integrated into relatively inexpensive ($600–$2,200) fishing and other Sonar systems that may be installed on a wide range of vessels. Even higher-resolution imagery is also available in systems ranging in cost from a few to several thousand dollars.

Forward-looking navigation Sonar systems are the equivalent of three-dimensional multibeam Sonars and can detect objects affixed to the bottom, exist within the water column, or lie in wait just on and below the surface at distances of 1,000 meters or more forward the bow (depending upon water depth). This includes uncharted rocks and reefs, shipping containers lost from ships, large marine mammals such as whales, buoys both on-station and adrift and even the hulls of other vessels ahead in the path of transit. At least one manufacturer (FarSounder, Inc., Warwick, Rhode Island) also provides a real-time bathymetry data overlay using live measurements onto navigation charts and captures bathymetric history as the vessel transits for future use [FarSounder 2019].

Visual and audible clues and sea state characteristics can provide experienced mariners present on the bridge with indications of imminent danger that may not be faithfully replicated at a remote control center. The use of side-scan and forward-looking navigation Sonar systems can result in extra margins of safety and reduce risk in MASS operations by determining the actual bottom contours and hazards that exist rather than the conditions present at the time of the last official survey. Their use can also provide an awareness of environmental changes along routes that would not otherwise be possible without the rapid availability of data at frequent intervals (each sailing) provided to the operators of MASS equipped with these Sonar systems.

These principles applied to MASS also are fully applicable to conventional vessels. The need for side-scan and forward-looking navigation Sonar as mandatory equipment onboard all vessels over 500GT should be considered in future IMO proceedings, and especially for updates to the Polar Code where unsurveyed waters and navigation charts lacking soundings are prevalent.

## 12.10 CROWDSOURCING MASS SUBSEA SENSOR DATA

With significant portions of the world's waterways, seas and oceans being poorly surveyed or having never been surveyed at all, the imagery available from side-scan Sonar and swath bathymetry from forward-looking Sonar can be useful to supplement national hydrographic agency assets in surveying these waters [Wright and Baldauf 2016a]. MASS can significantly increase the volume of high-resolution bathymetry and bottom imagery available over select routes by contributing to this effort [Wright and Zimmerman 2015].

Such an approach is already being implemented using single-beam echosounders, the primary method for hydrographic survey until the 1990s, when it was replaced by multibeam Sonar. Line soundings from single-beam echosounders are acquired by the National Oceanographic and Atmospheric Administration (NOAA) National Centers for Environmental Information (NCEI) under the auspices of the International Hydrographic Organization (IHO) Crowdsourced Bathymetry Working Group (CSBWG), who issued their draft Guidance on Crowdsourced Bathymetry in 2018 [B-22 2018]. Despite side-scan and multibeam Sonar providing the bulk of modern hydrographic survey data, the CSBWG effort is presently unable to accept other than single-beam echosounder data. IHO and their member states should consider the potential for expanding their crowdsourced data collection efforts to include side-scan imagery and bathymetry from forward-looking navigation Sonar. Side-scan Sonar can also provide information regarding fisheries and other environmental information useful for many purposes and scientific studies that span far beyond the needs of MASS [Wright 2017].

The availability of MASS-sourced high-resolution bottom imagery and bathymetry from each voyage on an ongoing and continuous basis can provide unprecedented insight into changes that occur in bottom topography over finite time periods and provide correlation to weather and other significant events. These data may also be used to create three-dimensional micro-spatial environmental corridor models representing the entire water column and its contents including fish, schools of fish, marine mammals and other aquatic life for environmental surveys and behavioral analysis. Sonar data may also extend to bottom composition determination, hazard to navigation detection and aid to navigation identification.

## 12.11 MASS WILL LEAD SHIPPING INTO THE FUTURE

The technological advances needed to implement MASS will pave the way to applying these same concepts across all shipping activities, manned and unmanned. Artificial-intelligence-based tools will enhance the ability of watchstanders to identify and avert potential collisions with other vessels and allisions with hazards to navigation that exist both above and below the waterline. The continuing automation of engine and auxiliary systems will lead the way in providing examples as to how this transformation will be accomplished. It will also provide unprecedented opportunities for gaining efficiencies in shipping operations on the vessels themselves, at shipping companies' headquarters and throughout the logistics chain. New communications technologies taking advantage of massive bandwidth compared to today's standards will provide much greater insight into heretofore unknown relationships between operational and maintenance data points to greatly enhance ship reliability and maintainability. The design features of MASS and its systems will, if properly implemented, provide greater immunity to physical and cyber threats and resilience in ship performance. Remote control center designs for MASS command, control and supervision will greatly expand the roles and capabilities of operations centers for all shipping enterprises.

Finally, the training and roles performed by licensed and unlicensed seafarers, and new roles performed by existing professionals untrained in maritime occupations yet essential to the operation of MASS can effectively transform the shipping industry as it is known today. The wealth of knowledge and experience that make up our maritime heritage and traditions can be preserved by maintaining existing command structures and continuing to add new roles to these structures much as has been done in the past. However, new opportunities exist in artificial-intelligence-based technologies to quantify and amass this knowledge to produce tools that can assist future seafarers to perform their jobs in a much safer and efficient manner.

## 12.12 POST-IMO REGULATORY SCOPING EXERCISE

The meeting in September 2019 of the IMO MASS Working Group for the Regulatory Scoping Exercise created recommendations considered by the Maritime Safety Committee at its meeting in May 2020 as to what maritime and related regulations are applicable to MASS. The next step is the determination of regulatory equivalences as provided for by the instruments or developing interpretations, amending of existing instruments, development of new instruments, and/or none of these choices as a result of the analysis. The scope of this undertaking is monumental and the details of how this will be accomplished will take many years to ascertain.

Meanwhile, test bed creation, prototype system development and MASS trials will continue to move forward, hopefully with the sharing of information between developers and regulatory authorities as anticipated in the Interim Guidelines for MASS Trials approved at the IMO MSC 101 meeting in June 2019. These results will be crucial for meaningful evaluation of not only MASS technologies but their regulation in both the short and long terms. How the issues of MASS competency and fitness for duty discussed earlier in this chapter are resolved is directly dependent upon the successful outcomes of these trials. However, the mechanisms for sharing trial results information are also yet to be determined. These are some of the more important issues that should be considered by the IMO through the early 2020s.

## REFERENCES

Abdollah, Tami. 2019. New Election Systems use Vulnerable Software. Associated Press, 13 July 2019. https://apnews.com/e5e070c31f3c497fa9e6875f426ccde1.

B-22 2018. Guidance on Crowdsourced Bathymetry. International Hydrographic Organization. Crowd Sourced Bathymetry Working Group (CSBWG). Edition 2.0.1, 2018. www.iho.int/iho_pubs/draft_pubs/B-12/CSB-Guidance_Document-Edition_2.0.1-Clean.pdf.

Efoui-Hess, Maxime. 2019. Climate Crisis: The Unstainable use of Online Video. The Shift Project, July 2019, p. 14.

FarSounder 2019. Sonar over Charts. www.farsounder.com/products/navigation_sonars.

Hughes, Chris. 2019. Iranian Bomb Boat Found in Path of British Warship HMS Duncan Sailing to Gulf. *The Daily Mirror*, 16 July 2019. https://www.mirror.co.uk/news/uk-news/british-warship-hms-duncan-en-18297526.

International Labour Office (ILO) 2011. Guidelines on the medical examinations of seafarersGeneva. ILO/IMO/JMS/2011/12. ISBN: 978-92-2-125096-8.

MSC 101/WP.8. Report of the Working Group (draft). Regulatory Scoping Exercise for the Use of Maritime Autonomous Surface Ships (MASS). Maritime Safety Committee. 101st session. Agenda item 5. 12 June 2019.

MSC 101/5/5. Interim Guidelines for MASS Trials. Maritime Safety Committee. 101st session. Agenda item 5/5. 8 April 2019.

Radiation Health Risks (RHR) 2019. Why 5G Cell Towers Are More Dangerous. https://www.radiationhealthrisks. com/5g-cell-towers-dangerous/.

SOLAS V. Safety of Life At Sea Convention. Chapter V, Regulation 20. Voyage Data Recorders.

STCW/CONF 2/34. Adoption of the Final Act and any Instruments, Resolutions and Recommendations Resulting from the Work of the Conference. Conference of the Parties to the International Convention on STCW. Attachment 2. 3 August 2010, p. 13.

TransNav 2019. First Session on MASS. *13th International Conference on Marine Navigation and Safety of Sea Transportation (TransNav)*. Gdynia, Poland, 12 June 2019.

Wakabayashi, Daisuke. 2019. Self-Driving Uber Car Kills Pedestrian in Arizona Where Robots Roam. *New York Times*, 19 March 2018. https://www.nytimes.com/2018/03/19/technology/uber-driverless-fatality.html.

World Health 2019. Thousands of Satellites Set To Launch For 5G. World Health Risks, 8 January 2019. https://www.worldhealth.net/news/thousands-satellites-set-launch-5g/.

Wright, R. Glenn. 2017. Scientific Data Acquisition using Navigation Sonar. *IEEE/ MTS Oceans Conference*, Anchorage, AK, September 2017.
Wright, R. Glenn. 2019. Enhanced MASS Situational Awareness using Virtual Navigation Aids. *Second International Conference on Maritime Autonomous Surface Ship (ICMASS 2019)*, Trondheim, Norway, November 2019.
Wright, R. Glenn and Michael Baldauf. 2014. Enhanced Situational Awareness through Muli-Sensor Integration. *Proceedings of 18th International Navigation Simulator Lecturers' Conference (INSLC 18)*, Buzzards Bay, MA, 2014. ISBN 978-0-692-29012-5, pp. 40–59.
Wright, R. Glenn and C. Zimmerman. 2015. Vector Data Extraction from Forward-Looking Sonar Imagery for Hydrographic Survey and Hazard to Navigation Detection. *Proceedings of IEEE/MTS Oceans Conference*, Washington, DC, 19–22 October 2015.
Wright, R. Glenn and Michael Baldauf. 2016a. Hydrographic Survey in Remote Regions: Using Vessels of Opportunity Equipped with 3-dimensional Forward-Looking Sonar. *Journal of Marine Geodesy*, 39, no. 6: 439–357. DOI: 10.1080/01490419.2016.1245266.
Wright, R. Glenn and Michael Baldauf. 2016b. Virtual Electronic Aids to Navigation for Remote and Ecologically Sensitive Regions. *Journal of Navigation*, 70, no. 2. DOI: 10.1017/S0373463316000527, pp. 225–41.

# Index

ABB, 205
AB-deck, 223
Able Seafarer, 185
Able Seaman, 185
Aboa Mare Maritime Academy, 183
Administrative support, 110
Afghanistan, 134
Aid to navigation (ATON), 105, 116, 117, 121, 123, 124, 125, 128, 133, 135, 136, 138, 139, 151, 240, 245, 246, 249
Air based sensors, 129
Aircraft carrier, 101
AIS aid to navigation (AIS-ATON), 116, 134, 135, 139, 246
AIS transponder buoys, 134
AIS vulnerabilities, 133
Alarm monitoring, 48
Alexa, 155
Allision, 123
Alstom Inspection Robotics, 74
Alternating current (AC), 45
Amazon, 157
American Bureau of Shipping (ABS), 13, 174, 200, 206
American Technology Corporation, 167
*Andrea Doria*, 6
An Industry Code of Practice, 197
Anti-submarine warfare Continuous Trail Unmanned Vessel (ACTUV), 11, 199
AP Moller-Maersk, 125, 173
Arctic Seas, 195
Area defenses, 165
Artificial intelligence (AI), 27, 49, 51, 52, 72, 94, 127, 137, 151, 153, 163, 180, 183, 187, 195, 240, 241

Asea Brown Boveri, 77
Assoc. for the Advancement of AI (AAAI), 52
Assured Autonomy (DARPA), 199
ASV Global, 9, 77
Atmospheric propagation, 134
ATON–manmade objects
 beacons, 116
 buildings, 116
 buoys, 116
 lighthouses, 116
 ranges, 116
 roads, 116
 towers, 116
ATON – natural features
 bluffs, 116
 islands, 116
 mountains, 116
 river mouths, 116
Attack, 169
Audio, 66, 88
Australia, 12, 199
Automated defensive measures, 168
Automated docking, 31
Automated loading and unloading, 81
Automated reasoning system, 155
Automated Ships Ltd, 11, 78
Automatic berthing, 47
Automatic berthing and unberthing, 81
Automatic Identification System (AIS), 6, 47, 48, 66, 71, 77, 78, 83, 103, 116, 117, 124, 130, 134, 138, 140, 146, 147, 153, 165, 172, 224, 242
Automatic locking mechanism, 168
Automatic Radar Plotting Aid (ARPA), 6, 47, 66, 103, 105, 124, 150
Automation, 37

255

Automation Officer, 109, 187
Autonomous aerial vehicle (AAV), 61, 68, 126, 158, 184, 188
Autonomous Spaceport Drone Ships, 79
Autonomous surface vehicle (ASV), 1, 158, 184, 188
Autonomous undersea vehicle (AUV), 60, 68, 128, 184, 188, 246
Autonomy, 37
Autopilot, 150, 175
Auxiliary systems, 46

Bachelor of Science, 181, 182
Backdoor vulnerabilities, 173
BAE Systems, 63, 167
Ballast, 46
Baltic and International Maritime Council (BIMCO), 13, 191, 200, 202
Barbed wire, 168
Battery, 61, 62, 94
Bay Ship and Yacht Co., 63
BCH Code, 235
Beamforming, 154
BeiDou, 129, 133
Belgium, 217, 233
Bell, 154
Bias, 51
Big Data, 129, 195, 245
BIMCO, 174
Black box, 88
Blockchain, 129, 245
Bluetooth, 187
Boeing, 136, 157, 243
Bonum Engineering and Consultancy, 63
Bore, 63
Bottom depth contours, 172
Bourbon, 11
Bridge to bridge communications, 146
Broadband, 48, 157
Broadband satellite communications, 69
Bulkhead, 220
Bunkering, 82
Buoy identification, 119
Bureau Veritas (BV), 13, 195, 200
Business Finland, 205

California State University Maritime Academy, 181
Canada, 12, 78, 79, 199
Captain, 91, 94, 97, 99, 108
Carbon footprint, 244
Cargotec, 205

Carpenter, 93
Casualty Investigation Code, 229
Caterpillar, 61
Caterpillar Marine, 73
Cavotec, 82
CEFOR, 204
Cellular, 130, 145, 156, 243
Cellular networks, 157
Central control station, 216
Central Intelligence Agency, 134
Central processing unit (CPU), 54
Certification, 149, 154
Chalmers Technical University, 11
Chief Engineer, 91, 94, 97, 101, 185
China, 5, 12, 19, 26, 75, 83, 193, 197, 215, 216, 217, 218, 219, 220, 221, 222, 223, 226, 228, 229, 230, 231, 232, 233, 234
China Classification Society (CCS), 13, 197, 201, 206
China Shipbuilding Industry Corporation, 206
China Ship Development and Design Center, 206
China State Shipbuilding Corporation, 206
Chlorine, 227
Chronometer, 5
*City of Racine*, 146
Classification societies, 12, 200
*Clermont*, 5
Closest point of approach (CPA), 124, 130, 140
Cloud computing, 112, 187
Cloud processing, 148
Coal heaver, 93
Code of Safety for Nuclear Merchant Ships, 219, 236
Collision avoidance, 117, 151
*Color Hybrid*, 62
COLREG, 192, 198, 203, 205, 223, 224, 225
COLREG Convention, 28, 47, 117, 140, 141, 149, 155, 182
COLREG rule 5, 117, 125
COLREG rule 7b,c, 117
Comité Maritime International (CMI), 13, 203, 204
Communication, 47, 216
Communications Display, 101, 105
Communications Officer, 92, 108, 186, 187

Company Security Officer, 161
Compass, 5
Compound annual growth rate, 23
Compressed air, 46
Compressed air gun, 167
Computer-aided design, 3, 4
Computer-aided engineering, 4
Computer-aided manufacturing, 4
Control Station, 216
CORE Advokatfirma, 195, 204
Corporate Representative, 97, 110
Cospas-Sarsat, 47, 146, 157
*Costa Concordia*, 50
Crew, 216
Crowd Sourced Bathymetry Working Group (CSBWG), 15, 128, 250
Crowd sourcing, 158
CSC Convention, 234
CSS Code, 231, 235
*C-Worker* 7, 78, 197
Cyber attack, 135, 173
Cyber security, 164, 170, 180, 183, 241
Cyprus, 222

Damage control, 48
Damen, 118
Danish Maritime Authority, 194, 195, 204
Database hacking, 135
Data breach, 173
Data mining, 158
Data monetization, 27
Data transfer rates, 148
Date, 88
Daylight, low light imaging, 66
Daylight camera, 124
Deck Department, 94, 111
Deck hand, 93
Deck officer, 184
Deep learning, 54, 137, 153, 240
Degrees of Autonomy, 90
Denial of service, 136
Denial of service attack, 136, 141, 172
Denmark, 12, 80, 90, 125, 194, 195, 199, 218
Depth finder, 68
Depth Sonar, 47
Depthsounder, 103
Det Norske Veritas, 201
Det Norske Veritas-Germanischer Lloyd, 13, 196, 197, 201
Diesel, 5, 61, 69

Diesel-electric, 61, 69
Digitalization, 13, 97, 173, 181, 195, 243
Digital Select Calling (DSC), 47, 155
Digital twin simulation, 51, 71, 108, 112, 113, 141, 185, 187, 197, 241, 248
DIMECC, 195, 205
Direct current (DC), 45
Distress, 154, 218, 224, 225
Doppler effect, 124
Doppler Radar, 132
Doppler shift, 125
Double bottom, 220
Dual fuel, 61
Dutch Ministry of Economic Affairs and Climate Policy, 196
Dynamic positioning, 47

*East Goodwin Sands Lightship*, 146
e-certificate, 214
Echosounder, 68, 88, 140
Eidesvik, 62
*Elbe*, 146
Electric, 5, 9
Electric light, 6
Electric wire, 168
Electromagnetic imaging, 65
Electromagnets, 81
Electronic Chart Display Information System (ECDIS), 6, 47, 88, 99, 103, 105, 116, 117, 124, 126, 133, 135, 138, 139, 147, 172, 175, 183
Electronic navigation chart (ENC), 6, 105, 124, 135, 138, 139, 150
*El Faro*, 50
eLoran, 32, 126, 138, 139, 140, 249
Emergency Position Indicating Radio Beacon (EPIRB), 47, 156, 217
e-Navigation, 145, 146
Enclosed spaces, 30
Engineering Department, 93, 101, 111
Engineering officer, 184, 185
Engineering systems, 150
Engineering work station, 101, 106
Engine operation, 65
Engines, 61
Eniram, 88
ENOVA (Norway), 74
Environment, 32
Environmental sensor systems, 118
  air, 118
  space, 118

subsea, 118
water surface, 118
Environment visualization, 64
Ericsson, 205
ESP Code, 228
Estonia, 12, 199
Ethernet, 150
Ethiopian Airlines Flight 302, 12
European Commission, 11, 194
European Research Commission, 194
European Union, 12, 193, 194
Executive Officer, 91, 97
Exterior defenses, 167
*Exxon Valdez*, 50

Facebook, 169
Facial recognition, 168
Falco, 76
FarSounder, Inc., 128, 249
Fault condition, 48
Federation of Norwegian Industries, 196
Ferries, 76
Finferries, 76
Fingerprint scan, 168
Finland, 12, 26, 76, 77, 88, 176, 182, 191, 193, 195, 197, 199, 220, 221, 226, 228, 229, 234
Finnish Funding Agency for Technology and Innovation, 195
Finnpilot Pilotage, 205
Fire extinguishing systems, 229
FireEye, 173
Firefighting, 9, 12, 29, 31, 59, 77, 78, 182, 216, 217, 230
Fire hose, 167
Fire safety, 216
First Mate, 91, 97, 99, 108
5G, 48, 151, 157, 243, 244
5G satellite, 243
5G terrestrial, 243
Fixed costs, 24
FlowChange, 75
*Folgefonn*, 31, 76, 82
Forward-Looking Infrared (FLIR), 78
Foul smelling liquid, 167
4G LTE, 157, 159
FTP Code, 233
France, 12, 199, 215, 225
*frequency* domain, 54, 122, 131, 133
Fresh water, 46
FSS Code, 228, 229
Fuel and power sources, 62
Fuel systems, 46

Fundamental tone, 154

Galileo, 129, 133
Gdansk, 26
General radio communications, 146
Germanischer Lloyd, 201
Germany, 82
Global Maritime Distress Safety System (GMDSS), 7, 47, 92, 146, 147, 156, 217
GlobalMET, 223
Global Navigation Satellite System (GNSS), 8, 32, 47, 62, 68, 71, 76, 82, 116, 117, 122, 126, 129, 130, 133, 134, 136, 138, 139, 140, 147, 172, 242, 245, 249
GLONASS, 129, 133
GNSS, 71, 138
GNSS signal outages, 133
GNSS vulnerabilities, 133
Gong, 154
Good seamanship, 224
Google, 11
GPS, 126, 129, 133
Graphics processing unit (GPU), 54, 56
Gray water, 46
*Great Intelligence*, 75
Green ammonia, 5, 63
Greenhouse gas, 244
Grounding, 123, 133
Guidance for Autonomous Ships, 202
Guidance on Crowdsourced Bathymetry, 250
Guidelines for Autonomous Shipping, 200

Harbor approaches, 83
Hazard to navigation, 105, 158
Heading, 88
Hearing, 117
Heat exchanger, 46
Heating and air conditioning, 46
Helmsman, 97, 99, 101, 103, 111
Helsinki City Transport, 77
High frequency (HF), 7, 47, 88, 146, 147, 149, 155, 187
Highly elliptical orbit, 157
Highly reliable systems, 69
High speed craft, 224
Hijacking, 133, 169, 172
Hike Metal, 78
HNA Technology & Logistics Group, 201, 206

Index 259

Horizon 2020, 194
Horten (NO) test area, 196
Hovercraft, 235
*Hrönn*, 10, 11, 78
HSC Code, 220, 235
Hudong-Zhonghua Shipbuilding, 206
Hull and deck design, *60*
HullBUG, 73
Human capabilities, 51
Human education and training, 51
Human error, 50
Human influence, 51
Human-machine interaction, 183
Human reasoning, 50
Human reasoning inaccuracy, 51
Human vision, 117
Hybrid systems, 61
Hydrofoil boats, 235
Hydrogen, 1, 5, 62, 63
Hydrogen fuel cell, 94
HYON, 75
Hyundai Heavy Industries, 62

IBC Code, 226
IEEE 802.11, 152
IGC Code, 227
IGF Code, 235
III Code, 233
IMDG Code, 226
IMO Degree Four Autonomy, 39, 42, 43, 49, 90, 95, 108, 111, 184, 215, 216, 217, 218, 223, 224, 225, 226, 229, 230, 233, 240, 248
IMO Degree One Autonomy, 39, 42, 90, 214, 215, 221, 223, 225, 226, 232
IMO Degree Three Autonomy, 39, 42, 43, 49, 90, 103, 108, 111, 184, 214, 216, 217, 218, 223, 224, 225, 226, 229, 230, 231, 248
IMO Degree Two Autonomy, 39, 42, 90, 214, 216, 223, 225, 226, 232
IMO MASS Working Group, 251
IMSBC Code, 230
Industry Project Autonomous Shipping, 196
Inertial navigation, 32, 140
Inertial navigation system (INS), 126, 136, 139, 150
INF Code, 228

Infrared, 75, 140
Infrared camera, 124
Infrared imaging, 65, 66, 139
INMARSAT, 47
Inmarsat, 75, 146, 157, 205, 243
INS, 138
Inspection Capabilities for Enhanced Ship Safety, 74
Institute of Marine Engineering, Science & Technology (IMarEST), 13, 21, 204
Integrated bridge system (IBS), 46, 149
Integrated Infrastructure Solutions, 82
Integrated navigation system, 149
Intercessional MSC Working Group, 214
Interim Guidelines for MASS Trials, 193, 239, 251
Intermodal shipping, 245
International Association of Classification Societies (IACS), 13, 201, 216, 228
International Association of Marine Aids to Navigation and Lighthouse Authorities (IALA), 55, 134
International Atomic Energy Agency (IAEA) Regulations, 228
International Chamber of Shipping (ICS), 13, 16, 203
International Federation of Shipmasters' Associations (IFSMA), 13, 203
International Grain Code, 230
International Group of Protection and Indemnity Clubs (IGP&I), 13, 204
International Hydrographic Organization (IHO), 15, 128
International Marine Contractors Association (IMCA), 13, 204
International Maritime Organization (IMO), xvii, 1, 9, 12, 14, 28, 37, 38, 43, 43, 49, 61, 62, 68, 83, 117, 123, 124, 126, 127, 128, 136, 138, 146, 154, 161, 174, 183, 189, 239, 247, 249, 252
International Organization for Standardization (ISO), 13, 191, 204
International Transport Workers' Federation (ITF), 13, 200, 203, 205

Internet, 173
Internet of Things (IoT), 26, 27, 80, 107, 108, 129, 148, 173, 174, 205, 245
Iran, 134, 215
Iris scan, 168
IS Code, 233
ISM Code, 219, 222
ISO Technical Committee 8, 205
ISPS Code, 161, 163, 170, 176, 220, 221

Jaakomeri (FL) test area, 195
Jamming, 134
Japan, 12, 193, 198, 199, 216, 217, 218, 219, 221, 222, 223, 226, 227, 228, 230, 231, 233, 234
Job safety, 29
Joint Industry Project Autonomous Shipping, 118
*Just Read the Instructions*, 79

Keppel Offshore & Marine, 80, 197
Keppel Singmarine, 80
Kongsberg, 9, 10, 11, 46, 74, 75, 79, 196, 205
Korea, 134, 192
Korean Register of Shipping (KR), 13, 201
*Kroonborg*, 73

L3 Technologies, 78
Latency, 43, 157
Law enforcement, 9
Lead-acid, 70
Lead-calcium, 70
Leaded line, 5
Legal instruments, 213
LeoSat, 157
Lethal force, 169
Liberia, 12, 197, 232, 233, 234
Lidar, 66, 75, 125, 132
Life boat, 217
Life jacket, 217
Life raft, 216
Life raft crew, 216, 220
Lighting, 46
Lights, 152
Light signals, 152
Lion Air Flight 610, 12
Liquefaction, 230
Liquefied natural gas (LNG), 1, 5, 62
Liquid natural gas, 94

Lithium-ion, 70, 82
Live imagery, 151
LL Convention, 191, 232
Lloyds Register, 13, 38, 62, 75, 195, 202
Local area network (LAN), 150
Long blast, 154
Long Range Acoustic Device, 167
Long Range Identification and Tracking (LRIT), 131, 146
Loran, 126
Low earth orbit, 157
Low light camera, 124
LSA Code, 217, 233, 237

Machine learning, 54, 137, 153, 163, 183, 187
Machine vision, 245
Maersk, 27, 63
*Maersk Pelican*, 63
Maintenance, 65
Maintenance inspections, 151
Malware, 135, 170
Mampaey Offshore Industries, 82
MAN, 61
Management Level, 91, 97, 99, 110
Maneuvering Characteristics Augmentation System (MCAS), 136
Manual fire-fighting equipment, 217
Marconi, 145
Marine Autonomy Research Site (MARS), 199, 206
Marine Conservation Zone, 32
Marine Industries Alliance, 195
Marine Inspection Robotic Assistant System, 74
Marine mammals, 123
Marine Protected Areas (MPA), 32
Marine Sanctuary, 32
Marin Teknikk, 74
Marinus of Tyre, 5
Maritime 2050, 196
Maritime academies, 180
Maritime Drone Operator, 188
Maritime incubators, 26
Maritime institutes, 180
Maritime Port Authority of Singapore, 197, 201
Maritime Safety Committee (MSC), 12, 38, 183, 189, 192, 213, 239, 251
Maritime safety information, 146

Index  261

Maritime Transportation Security Act, 170
Maritime UK, 41
Maritime Unmanned Navigation through Intelligence in Networks (MUNIN), 11, 24, 89, 95, 96, 111, 194
MARPOL Convention, 203, 213, 235
Marshall Islands, 223
MASS reasoning, 49
Massterly, 75
Master, 185, 216, 230, 233
Master of Engineering, 182
Master's degree, 181
Mate, 185
*Mayflower*, 11
McDonough Marine Service, 79
Mediterranean Sea, 134
Medium frequency (MF), 47, 146, 149, 155
Metadata, 119
Metal Shark, 9, 77, 78
Methanol, 62
Metrics of Autonomy
   (IMO) Degree Four, 42
   (IMO) Degree One, 42
   (IMO) Degree Three, 42
   (IMO) Degree Two, 42
   (LR) Autonomy Level 0, 38
   (LR) Autonomy Level 1, 38
   (LR) Autonomy Level 2, 38
   (LR) Autonomy Level 3, 38
   (LR) Autonomy Level 4, 38
   (LR) Autonomy Level 5, 38
   (LR) Autonomy Level 6, 38
   (MUK) Level of Control 0, 41
   (MUK) Level of Control 1, 41
   (MUK) Level of Control 2, 41
   (MUK) Level of Control 3, 41
   (MUK) Level of Control 4, 41
   (MUK) Level of Control 5, 41
   (NFAS) Automatic, 40
   (NFAS) Constrained Autonomous, 41
   (NFAS) Decision Support, 40
   (NFAS) Fully Autonomous, 41
   (NTNU) Level of Autonomy 1, 40
   (NTNU) Level of Autonomy 2, 40
   (NTNU) Level of Autonomy 3, 40
   (NTNU) Level of Autonomy 4, 40
   (SAE) Level 0 – No Automation, 42
   (SAE) Level 1 – Driver Assistance, 42
   (SAE) Level 2 – Partial Automation, 42
   (SAE) Level 3 – Conditional Automation, 42
   (SAE) Level 4 – High Automation, 42
   (SAE) Level 5 – Full Automation, 42
Metrics of autonomy, 37
M/F *Teistin*, 63
Michigan Technological University, 199
Microphone arrays, 154
Microphones, 152
Microwave, 145, 156, 157, 158
Military, 9
Mine, 169
Ministry of Defence (NL), 196
Ministry of Infrastructure and Water Management (NL), 196
Ministry of Land, Infrastructure, Transportation and Tourism (JP), 198
*Minnehaha*, 146
Mitsui OSK Lines, 198
MmRadar, 66, 125, 131, 132, 138
Mob Excess Deterrent Using Silent Audio, 167
Monohakobi Technology Institute, 205
Mooring Systems, 81
Morse Code, 88
Multibeam Sonar, 249
Multiple modes of autonomy
   conventional manned, 9
   reduced manned, 9
   unmanned, 9
Multiple redundant systems, 70
Multiple sensor modalities, 118
Mumbai, 26

National Centers for Environmental Information (NCEI), 250
National Electronics Manufacturers Association (NEMA), 8
National Oceanographic and Atmospheric Administration (NOAA), 135
National Timing Security and Resilience Act, 126
Nautical chart, 5
Nautical Institute (NI), 13, 205
Naval, 9
Naval Ship Code, 202
Navigation, 5, 46, 65
Navigational telex (NAVTEX), 47, 146, 156
Navigation chart, 6, 15, 27, 105, 124, 128, 133, 150

Navigation Display, 103
Navigation reasoning system, 247
Navigation Sonar, 32, 68, 71, 103, 116, 127, 128, 132, 136, 139, 150, 158, 248, 249, 250
Netherlands, 79, 81, 82, 83, 196, 199, 217, 233
Neural network, 52, 163, 240
　hidden layer, 52
　input layer, 52
　interconnections, 52
　output layer, 52
Neural processing unit (NPU), 56
Neuron, 52
New Zealand, 81, 231
Nickel-cadmium, 70
Nigeria, 219, 222
Nippon Yusen, 198
NMEA 0183, 64, 70, 175
NMEA 2000, 64, 70, 150
Non-Governmental Organization, 13, 202
Non-lethal defenses, 168
Nordic Association of Marine Insurers, 195
Norled, 76
North America, 26
Norway, 9, 12, 26, 63, 76, 191, 193, 196, 199, 219, 222
Norwegian Coastal Administration, 196
Norwegian Coastal Administration VTS, 75
Norwegian Defence Research Establishment (FFI), 196
Norwegian Forum for Autonomous Ships (NFAS), 40, 195, 196
Norwegian Marine Technology Research Institute, 196
Norwegian Maritime Authority, 196
Norwegian University of Science and Technology (NTNU), 196
NotPetya malware, 174
Novia University, 182
Nuclear, 1, 5, 21, 22, 228, 236
Nuclear powered ships, 219
NVIDIA, 56

Oars, 2
Oarsman, 93
Obsolete software, 170
*Of Course I Still Love You*, 79
Officer In Charge of a Navigation Watch, 185
*Ohio*, 146
One Sea Autonomous Maritime Ecosystem, 13, 195, 205
OneWeb, 157, 243
Operating costs, 25
Operational Level, 92, 99, 103
Operations center, 187
Optimization processing unit (OPU), 56
Out-of-tolerance condition, 48

Pacific Maritime Institute, 11
PCI eXtensions for Instrumentation, 70
Pennants, 5
Perimeter defenses, 166, 167
Perimeter Sonar, 68, 128
Personal credentials, 168
Personal watercraft, 165
Physical constraint, 168
Physical security, 164
Pigeoneer, 93
Pilotage, 138
*Pixel* domain, 131
Polar Code, 221, 226, 249
Polar Water Operational Manual, 221
Portable extinguishers, 217
Port and harbor facilities, 81
Port of New York and New Jersey, 173
Position, 88
Precision navigation and timing (PNT), 126, 129, 130, 136
Predictive maintenance, 72
Process automation, 44, 49
Promare, 11
Propagation delay, 43
Propeller pitch, 88
Proper lookout, 223, 224
Propulsion and power generation, 61
Propulsion control, 44
Public correspondence traffic, 146

Radar, 6, 8, 18, 47, 66, 71, 75, 77, 88, 99, 103, 105, 116, 117, 123, 126, 131, 138, 140, 151, 153, 165, 172, 242, 246
Radar beacon, 6
Radio, 88
Radio frequency, 243
*RALamander*, 197

Ramboll-Core, 195
*R Amora*, 10, 11, 79
Rating Forming Part of a Navigational Watch, 185
Raymarine, 82
Raytheon, 167
Razor wire, 168
Reasoning, 37
Recon Robotics, 74
Recon Scout, 74
Red Sea, 134
Regulatory scoping exercise, 12, 118
Remote aerial vehicle (RAV), 61, 68, 126,129, 158, 184, 188
Remote control, 216, 217, 218, 224, 229, 232
Remote control center (RCC), 40, 41, 49, 51, 87, 89–93, 95, 96, 97, 99, 101, 103, 107–113, 144–149, 152, 155, 161, 162, 164–170, 172, 173, 174, 179, 181, 184–187, 216, 242
Remote operator, 222, 223, 224, 229, 232, 233
Remote surface vehicle (RSV), 158, 184, 188
Remote undersea vehicle (RUV), 60, 68, 128, 184, 188
*Rena*, 50
*Republic*, 146
Republic of Korea, 126, 192, 200, 219, 222
Rescue, 9
Rescue boat, 217
Responsible person, 216
Revolving light, 155
*R F Matthews*, 146
Rigger, 93
Robert Allan Ltd, 10, 11, 79
Robert Fulton, 5
RoboShip, 73
Robotic devices, 69, 94
Robotic maintenance, 73
Robotic systems, 151
Rocket propelled grenade, 169
RO Code, 234
Rolls-Royce, 11, 76, 80, 96, 97, 111, 195, 206
Rotterdam, 26
Royal Wagenborg, 73
Rudder order, 88
Russian Federation, 219, 222

Safety, 28
Safety Center, 216
Safety driver, 247
Safety of navigation, 28, 180
SailRouter B.V, 27
Sails, 2, 4
Samskip, 75
San Francisco Bay, 63
SAR Convention, 203, 225
Satellite, 48, 145, 157
Satellite AIS, 6
Satellite networks, 187
SCC Functional Unit, 111
SCC Security Officer, 93, 107
Screen and Vessel Controller, 106
Scuba diver, 169
Seabed contour following, 32
Seafarer, 214, 215, 216, 217, 220, 223, 224, 226, 227, 229, 230, 231, 232
*Sea Hunter*, 11, 199
Sea Machines, 11, 78, 125
Search and rescue, 77, 225
Search and Rescue Transponder (SART), 48, 156, 217
Searchlight, 154
Sea service requirements, 185, 188
Seashuttle project, 62, 75
SeaZip, 79, 118
*SeaZip 3*, 79, 118
Second Engineer, 99, 101, 103, 106, 185
Second Mate, 97, 103
Securing with wire mesh, 231
Security, 44
Security Awareness, 162
Semaphore, 6
Sensor placement, 65
Sensors, 64
Sensor suite composition, 65
Service-Level Agreement, 171
Service requirements, 187
Seventh Framework Programme, 194
Sewage, 46
*Sewol*, 50
Sextant, 5
Shapes, 152
Sharktech, 9, 10, 77
Sheridan levels of automation, 182
Shipboard Autonomous Fire Fighting Robot, 73
Shipboard sensors, 66, 120
Ship Inspecting Robot, 74

Ship operations and monitoring center, 89
Ship Security Alerting System (SSAS), 146
Shore control center organization, 110
Short blast, 154
*A Shortfall of Gravitas*, 79
Side-scan Sonar, 68, 71, 120, 122, 128, 132, 139, 158, 248, 249, 250
Signal flags, 5
Silent Guardian, 167
Simulators, 182
Singapore, 12, 26, 83, 193, 197, 199, 218, 223
Singapore Maritime Port Authority, 80
Single beam echosounder, 7, 68, 88, 116, 117, 122, 127, 128, 136, 139, 140, 150, 246, 250
Single point of failure, 70, 133, 150
SINTEF Ocean, 196
Sir George Biddell Airy, 2
Siri, 155
Sites of Specific Scientific Interest (SSSI), 32
Situational awareness, 65, 125, 248
*Slavonia*, 146
Slippery foam, 167
Small craft, 165
Small vessel identification, 119
Smart chair, 97
SMArt Maintenance of Ships, 73
Smart ports, 244, 245
Smart sensors, 164
Smart Ships Coalition (SSC), 13, 206
Society of Automotive Engineers (SAE), 42
Software simulation, 54
SOLAS Convention, 47, 146, 148, 156, 162, 163, 191, 198, 202, 203, 213, 214, 215, 219, 221, 223, 226, 227, 228, 231
Sonar, 83, 131, 139, 151, 169, 246
Sound monitoring, 124
Sound signals, 152
South Africa, 12, 199
South Korea, 126
Space based sensors, 129
Space STP Protocol, 236
SpaceX, 79, 157, 243
Spain, 222, 223, 225, 231
Special Areas of Conservation (SAC), 32
Special vessel, 153

Speed, 88
Sperry, 46
Spire Global, 243
Spoofing, 126, 133, 134, 136, 172
Spyware, 169
Statistical analytics, 139
Statistical methods, 72
Statistical processes, 153
STCW Convention, 153, 162, 180, 182, 183, 184, 185, 198, 203, 213, 222, 223, 231, 240
STCW-F Convention, 231
Steam engine, 5
*Stockholm*, 6
Storfjorden (NO) test area, 196
STP Agreement, 236
Strandfaraskip Landsins, 63
Strapping or lashing, 231
Strobe light, 155, 168
SUA Convention, 203
Subsea sensors, 68, 78, 118, 121, 122, 127, 139, 248, 250
*Suomenlinna II*, 77
Supervised learning, 54, 55
Support Level, 93, 109
Surface sensors, 123
Surveillance, 9, 77
Survey, 9, 77
Svitzer, 80
*Svitzer Hermod*, 80
Sweden, 11, 12, 199, 215, 223
Switzerland, 82
System electronic navigation chart (SENC), 6
Systems Display, 101, 106

Tear gas, 168
Technical experts, 99, 109, 112
Technical University of Denmark, 194
Technology Centre for Offshore and Marine Singapore, 80, 197
Tekes, 195
Telegraph, 6
Teleprinter exchange (telex), 88
Test bed, xvii, 83, 251
Third Engineer, 99, 101, 103, 106
Third Mate, 97, 99, 103, 182
Threat assessment, 165
3-D micro-spatial environmental corridor, 250
3G, 150
Thrustmaster, 79

Tieto, 205
Time, 88
*Time* domain, 131
*Titanic*, 146
TKI-Maritiem, 79, 196
Tonnage Convention, 234
Transportation Worker Identification Card, 170
Trondheim Port Authority, 196
Trondheimsfjord (NO) test area, 196
Trusted personnel, 169
Turbine, 1, 5
Turkey, 199, 217, 225
Twitter, 169
Type 1G ship, 227
Type of propulsion, 153

UAV, 126, 133
UAV/UUV Operator, 109
Uber, 246
UK Maritime, 197
Ukraine, 174
Ulstein, 62
Ultra high frequency (UHF), 187
Unauthorized entry, 164
Under keel clearance, 88
United Arab Emirates, 200
United Kingdom, 12, 78, 83, 193, 196, 199
United States, 12, 77, 78, 79, 125, 126, 128, 134, 170, 181, 198, 199, 206, 222, 223, 238, 242, 249
Universities, 180
University College Cork, 11
University College of South East Norway, 196
University of Michigan, 199
University of Texas at Austin, 133
University of Twente, 73
Unlicensed ratings, 185
Unmanned aerial vehicle, 129
Unmanned Cargo Ship Development Alliance (UCSDA), 12, 201, 200, 206
Unmanned Marine Systems Code, 202
U.S. Coast Guard, 30, 133, 135, 162, 175
U.S. Defense Advanced Research Projects Agency (DARPA), 199
U.S. Maritime Administration, 11
U.S. National Bureau of Standards (NIST), 174
U.S. Naval Research Laboratory, 73

U.S. Navy Office of Naval Research, 73
U.S. Office of Naval Research, 11, 199

Vacuum pods, 81
Validation, 65
Vard, 74
Velodyne, 125
Verification, 65
Very high frequency (VHF), 47, 99, 105, 124, 130, 134, 146, 147, 149, 152, 153, 155, 158, 187
VesselBot, 27
Vessel Personnel with Designated Security Duties, 162
Vessel Safety Assessment, 162
Vessel Security Officer (VSO), 93, 107, 108, 162, 184, 186
Vessel Security Plan (VSP), 161, 163, 171, 172
Vessel Traffic Service Portfolio, 147
Vessel Traffic Services (VTS), 48, 155
VHF radio, 246
Video, 66, 80
Video imagery, 158
*Viking Energy*, 62
Virginia Tech, 73
Virtual aid to navigation (VATON), 32, 83, 116, 117, 121, 246
Virtual bridge, 95
Visual imaging, 65
Visual monitors, 31
Visual sensors, 152
Voith Schneider, 79
Voyage data recorder (VDR), 71, 88, 126
*Vulcan*, 3

Wärtsilä, 31, 46, 61, 76, 205, 206
Watchkeeping, 65, 223, 231
Water cannon, 167
Water fire-extinguishing system, 227
*Water-Go-Round*, 63
Water-spray system, 219
Wavefoil, 63
Weather Radar, 99
Whales, 123
Whistle, 154
Wide area network (WAN), 48, 123, 156, 157, 158, 187

Wi-Fi, 130, 152, 180, 187, 246
Wind, 94
Wind direction, 88
Windows 7, 170
Windows NT, 170, 175
Windows XP, 170
Wind power, 4
Wind speed, 88
Working Group on MASS, 12, 189, 191, 198, 199, 204
Wuhan University of Technology, 197

X-band Radar, 48, 156

Yanmar, 61
*Yara Birkeland*, 9, 10, 33, 62, 74, 196, 244
Yara International ASA, 74

Zhuhai, Guangdong (CN) test area, 197, 201
Zhuhai Yunzhou Intelligence Technology, 197